Best Crop Management and Processing Practices for Sustainable Cotton Production

Edited by Songül Gürsoy and Songül Akın

Published in London, United Kingdom

Best Crop Management and Processing Practices for Sustainable Cotton Production
http://dx.doi.org/10.5772/intechopen.107795
Edited by Songül Gürsoy and Songül Akın

Contributors
Abid Ali, Ahmad Nawaz, Aiman Khalid, Amit Madhu, Ashfaq Ahmad Rahi, Deqiang Qin, Guoxing Wu, Habib Ali, Hasan Taha, Iqra Maryum, K. Raja Reddy, Mehboob Hussain, Mirza Abdul Qayyum, Muhammad Naeem Akhtar, Muhammad Waseem Akhtar, Muhammad Yasin, Natraj Krishnan, Shafqat Saeed, Shien Lu, Songül Akın, Songül Gürsoy, Tanveer ul Haq, Te-Ming Tseng, Umer Sharif, Unsar Naeem-Ullah, Wania Hasan Zubairi, Waqas Wakil, Xiaoping Qin, Xi Gao, Zeeshan Shahid, Ziming Yue

First published in London, United Kingdom, 2024 by IntechOpen
IntechOpen is the global imprint of INTECHOPEN LIMITED, registered in England and Wales, registration number: 11086078, 167-169 Great Portland Street, London, W1W 5PF, United Kingdom

British Library Cataloguing-in-Publication Data
A catalogue record for this book is available from the British Library

Additional hard and PDF copies can be obtained from orders@intechopen.com

Best Crop Management and Processing Practices for Sustainable Cotton Production
Edited by Songül Gürsoy and Songül Akın
p. cm.

This title is part of the Agricultural Sciences Book Series, Volume 9
Topic: Agronomy and Horticulture
Series Editor: W. James Grichar
Topic Editor: Ibrahim Kahramanoglu
Associate Topic Editors: Murat Helvaci and Olga Panfilova

Print ISBN 978-0-85014-009-5
Online ISBN 978-0-85014-010-1
eBook (PDF) ISBN 978-0-85014-011-8
ISSN 3029-052X

For EU product safety concerns:
IN TECH d.o.o., Prolaz Marije Krucifikse Kozulić 3, 51000 Rijeka, Croatia,
info@intechopen.com or visit our website at intechopen.com.

IntechOpen Book Series
Agricultural Sciences
Volume 9

Aims and Scope of the Series

The importance of agriculture cannot be overstated. It helps sustain life, as it gives us the food we need to survive and provides opportunities for economic well-being. Agriculture helps people prosper around the world and combines the creativity, imagination, and skill involved in planting crops and raising animals with modern production methods and new technologies. This series includes two main topics: Agronomy and Horticulture, and Animal Farming. This series will help readers better understand the intricacies of production agriculture and provide the new knowledge that is required to be successful. The success of a farmer in modern agriculture requires knowledge of events happening locally as well as globally that impact input decisions and ultimately determine net profit.

Meet the Series Editor

W. James Grichar has been employed with Texas A&M AgriLife Research for over 45 years with an emphasis on research in agronomy, plant pathology, and weed science. He obtained his BS from Texas A&M in 1972 and his Masters of Plant Protection in 1975. He has published 195 journal articles, over 330 research reports and briefs, 11 book chapters, and over 300 abstracts of profession meetings. He also directs research in many crops including corn, grain sorghum, peanuts, and sesame. He has held various positions in different professional societies including the American Peanut Research and Education Society, Southern Weed Science Society, and Texas Plant Protection Conference in addition to being Associate Editor for Peanut Science and Weed Technology. Significant accomplishments have included spearheading efforts to determine the optimum planting time for soybean production along the upper Texas Gulf Coast. These efforts have shown growers that soybean yields can be improved by 10 to 20% by following a late March to early April plant date. He also has been instrumental in developing a herbicide program for peanut production in the south Texas growing region. Through the development and use of herbicides that are effective against major weed problems in the south Texas region, peanut yields have increased by 25 to 30%.

Meet the Volume Editors

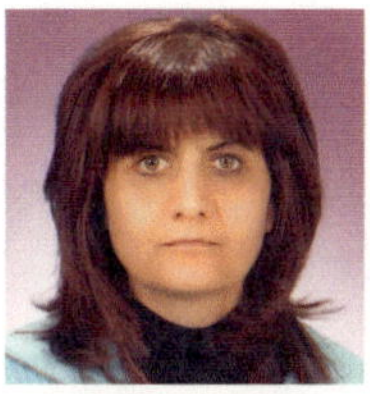

Dr. Songül Gürsoy is a professor in the Department of Agricultural Machinery and Technologies Engineering, Dicle University, Türkiye. She obtained a bachelor's degree in Agricultural Mechanization and an MSc in Agriculture Machinery from Atatürk University, Türkiye, in 1989 and 1997, respectively. She obtained a Ph.D. in Agriculture Machinery from Cukurova University, Türkiye, in 2005. She completed her postdoctoral studies in the Department of Biosystems Engineering, University of Manitoba, Canada. Dr. Gürsoy has been researching and teaching in the field of agricultural engineering, focusing on the performance of tillage and planting machinery, mechanical weed management, processing technologies of field crops, drying technology, and precision agriculture. She has been conducting many projects on the conservation of tillage systems and mechanical weed control in the production of field crops. She has more than sixty scientific publications to her credit.

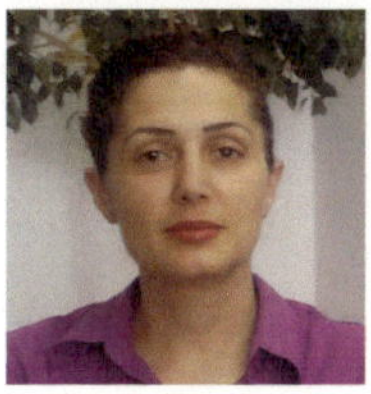

Dr. Songül Akın is an associate professor in the Department of Agriculture Economy, Dicle University, Türkiye. She has been researching and teaching in the field of agricultural extinction and economy. She is the author of more than forty journal and conference articles.

Contents

Preface

Cotton, an arable crop used mainly for fiber, plays a dominant role in the industrial and agricultural economy of the world. The cottonseed, which remains after cotton is ginned, is used to produce oil for human consumption and oilseed cake for animal feed. Nearly 35 million hectares are under cotton cultivation around the world, representing about 2.5% of the world's arable land. About 25 million tons of cotton lint is produced each year, of which nearly 40% is traded internationally. One of the most important problems in cotton production is the high consumption of pesticides compared to other products. It is estimated that 25% of the worldwide use of insecticide and 10% of pesticide use are accounted for by cotton cultivation. Pesticides sprayed across cotton fields easily run off and pollute freshwater sources.

Crop management and processing in cotton production begins with seedbed preparation, sowing of seeds, and crop maintenance and ends with crop harvest, handling, storage, and marketing. The best crop management and processing practices improve crop productivity and can contribute to greater yields with improved quality. These practices allow diverse actors such as farmers, companies, government agencies, and nongovernmental organizations (NGOs) to optimize resource use efficiency, create marketable products, reduce waste, assure market access, and reduce the risk of adverse relations with local stakeholders. Hence, crop management and processing practices are important aspects to ensure optimal productivity and even agricultural production.

In recent years, sustainable crop production and management have received increased attention at the global level because the use of agrochemicals and mechanization in agricultural production is producing an intense ecological crisis worldwide. Sustainable crop management focuses on agricultural techniques that are sustained over a long period without damage to the environment, or with only minimal damage. Sustainable cotton production means using natural resources such as water, land, and energy more efficiently. It pioneers commonsense cotton production practices that significantly reduce water use, land use, soil loss, and energy use while improving soil health and yield per acre. The development of innovative technologies, management systems, and protective approaches guided by science-based environmental goals and objectives has been key to these advances in sustainable cotton production.

The key objective of this book is to raise awareness and improve irrigation, plant nutrition management, integrated pest management, fiber quality, and biodiversity in cotton production fields. It includes seven chapters organized into three sections: "Crop Management Practices in Cotton Production," "Pest Management Practices in Cotton Production," and "Sustainability and Precision Agriculture in Cotton Production."

Chapters in Section 1 include "Utilization of Secondary Metabolites in Cotton Production", "Enhancing Water Use Efficiency by Using Potassium-Efficient

Cotton Cultivars Based on Morphological and Biochemical Characteristic" and "Naturally Colored Cotton: A Sustainable Innovation". Section 2 includes "Role of Biotic and Abiotic Factors for Sustainable Cotton Production" and "Integrated Pest Management in Cotton". Section 3 includes "A Review of the Factors Affecting Adoption of Precision Agriculture Applications in Cotton Production" and "The Strategic Importance of Cotton Production for the World and Türkiye".

The editors would like to give special thanks to the authors for their outstanding and timely work in producing such fine chapters. We are also thankful to Publishing Process Manager Mr. Dominik Samardzija and the rest of the staff at IntechOpen for their help throughout the publication process. We hope that the readers of this book will benefit.

Songül Gürsoy
Department of Agricultural Machinery and Technologies Engineering,
Dicle University,
Diyarbakır, Türkiye

Songül Akın
Department of Agricultural Economy,
Dicle University,
Diyarbakır, Türkiye

Section 1

Crop Management Practices in Cotton Production

Chapter 1

Utilization of Secondary Metabolites in Cotton Production

Ziming Yue, Te-Ming Tseng, K. Raja Reddy, Natraj Krishnan and Shien Lu

Abstract

Cotton is the most critical fiber crop and one of the top three oilseed crops in the world. One pronounced feature of cotton is that it is rich in secondary metabolites, mainly including terpenoids, flavonoids, and phenolic acids. These secondary metabolites have various ecological roles, such as defense and signal transmission. With the concept of plant secondary metabolites becoming more and more evident in the mid-twentieth century, cotton secondary metabolites as natural phytoalexins were also established. Terpenoids are stored in pigment glands that are distributed almost all cotton plant surfaces or subsurfaces and defend cotton plants from chewing insects, pathogens, and other herbivores. Flavonoids are relevant to fiber quality and color and also play a role in mechanism in insect and pathogen resistance. Phenolic acids play a role in weed suppression and insect and pathogen resistance. There are several reviews on cotton secondary metabolites, and the most recent one was five years ago. They all focus on the metabolites themselves. None of them focus on applications in cotton production. This review started from browsing the abundant literature on cotton secondary metabolites, and then analyzing their potential application in cotton production. Finally, our recent findings were discussed in this chapter.

Keywords: cotton, terpenoid, flavonoid, phenolic acid, insect, pathogen, weed

1. Introduction

Cotton is the world's most important natural fiber crop and one of the top oilseed crops among soybean, rapeseed, canola, sunflower seed, and peanut. Another by-product is the protein-rich meal from cotton seeds. Cotton is grown in 75 countries worldwide [1]. Different from other crops, cultivated cotton includes four species in the genus *Gossypium*: *G. hursutum*, *G. barbadense*, *G. arboretum*, and *G. herbaceum*. The first two are allotetraploid and distributed in the New World (Americas), and the latter are diploid and distributed in the Old World (Asia and Africa). The genus *Gossypium* belongs to the *Malvaceae* family, and it has approximately 50 diploid species among eight genomes (A–G and K genomes) and seven allotetraploid species [2, 3].

Beyond its economic value, cotton also possesses a rich array of secondary metabolites and organic compounds synthesized by plants for various ecological functions.

These secondary metabolites play a crucial role in the plant's defense against pests, pathogens, and environmental stresses. There are at least three reviews on cotton secondary metabolites: [4–6]. Ref. [4] is comprehensive and expected the cotton secondary metabolites to be utilized to defend against pests in cotton production, but data accumulation was not as abundant as in the present day, and that was published before GMO cotton appearance; Ref. [5] focused more on cotton terpenoids; and Ref. [6] focused on cotton flavonoids. This review will follow the coverage of Ref. [4] and review the logical development of the utilization of secondary metabolites in cotton production.

In recent years, there has been a growing interest in harnessing the potential of cotton secondary metabolites for applications in agronomy. Part of the reason is that cotton is still the "dirtiest" crop as it consumes 10.24% of the insecticides and 4.71% of pesticides in 2019 with only 2.5% of the arable land according to the International Cotton Advisory Committee [7], and it was one of the early crops that widely adopted GMO technology [8].

The key distinction between the eras of Refs. [4, 5], and the current period lies in the acknowledgment of secondary metabolites in plants as an innovative foundation for potential biopesticides. This recognition has opened avenues for their application in sustainable agriculture, as highlighted by Ref. [9]. Researchers and agronomists have identified the diverse properties of secondary metabolites, including antimicrobial, insecticidal, antioxidant, and allelopathic activities. These bioactive compounds have the potential to revolutionize agricultural practices by offering environmentally friendly alternatives to synthetic pesticides, fertilizers, and growth regulators.

The use of cotton secondary metabolites in agronomy presents several advantages. Firstly, it provides an opportunity to reduce the reliance on conventional chemical inputs, thereby minimizing the environmental impact of their use. Secondly, these natural compounds may offer a sustainable solution to combat pests while maintaining crop productivity. Moreover, using cotton secondary metabolites can lead to the development of novel bio-based products and promote the concept of integrated pest management (IPM) and sustainable agriculture.

This review explores the current knowledge and research advancements in using cotton secondary metabolites in agronomy. It will delve into the compounds' chemical diversity, modes of action, and potential applications in crop protection, nutrient management, and plant growth promotion. Furthermore, the challenges and opportunities associated with their practical implementation will be discussed, along with future perspectives for utilizing cotton secondary metabolites in sustainable agricultural systems.

By critically evaluating the existing literature and highlighting key findings, this review intends to provide a comprehensive overview of the potential benefits and limitations of incorporating cotton secondary metabolites into agronomic practices. Ultimately, such insights can contribute to developing innovative strategies that harness the power of these bioactive compounds to promote sustainable and environmentally conscious approaches to crop production.

2. Secondary metabolites in cotton

As the name initially meant, plant secondary metabolites used to be considered as waste in metabolism without primary growth, development, and reproduction functions [10]. Nowadays, plant secondary metabolites usually refer to small molecules produced in metabolism with ecological functions, such as defense and survival [10]. Plant secondary metabolites have a long history of being used as herbal medicine; their use in agriculture is a recent development. They are classified into terpenoids, phenolic compounds (flavonoids and phenylpropanoids), and N-containing compounds (cyanogenic glycosides, alkaloids, and glucosinolates) based on their biosynthetic pathways [11].

2.1 Cotton secondary metabolites

Cotton secondary metabolites are flavonoids (including tannins), terpenoids, phenolic acids, and fatty acids. Alkaloids and sulfur-containing secondary metabolites have not been emphasized due to their number so far.

2.1.1 Flavanoids

Flavonoids have a backbone diphenylpropane structure (C_6-C_3-C_6) where two aromatic rings are linked *via* a three-carbon chain (**Figure 1**). The A ring is typically formed from a molecule of resorcinol or phloroglucinol synthesized *via* the acetate pathway with a characteristic hydroxylation pattern at positions 5 and 7 [12]. The B ring comes from the shikimate pathway with a characteristic hydroxylation pattern at positions 4′-, 3′, 4′-, or 3′, 4′, 5′-. One impressive phenomenon is that the number of characterized cotton flavonoids has significantly increased from 23 in 1994 [13] to 52 in 2017 [6], and up to 190 in 2023 [14]. With the MS/MS technique, it is easier to

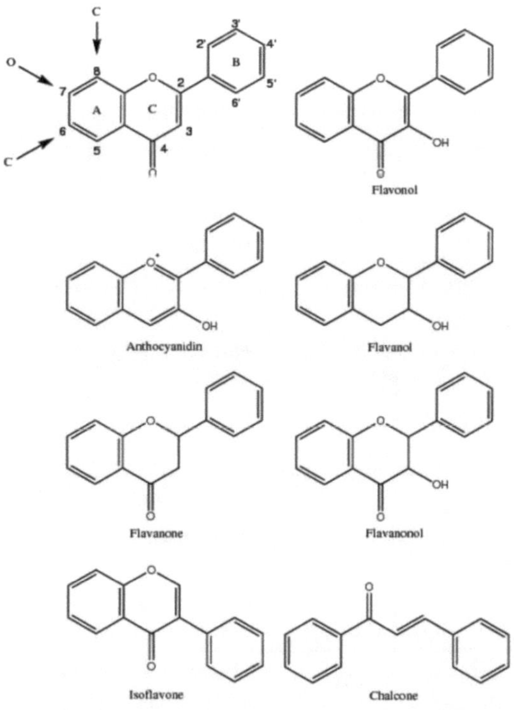

Figure 1.
Typical flavonoid structures.

identify more and more flavonoids, while the functions of the newly discovered flavonoids in cotton plants largely remain to be investigated. Of the reported ecological functions, signaling, defense, and coloring functions have been well studied as listed in **Table 1**.

Name	Structure	Plant/part	Function	References
Trifolin	kaempferol-7-O-β-D-galactoside	Flowers from *G.* sp.		[15]
Isoastragalin	kaempferol-3-O-α-D-glucofuranoside	Flower from *G. hirsutum*		[13, 15–17]
Nicotiflorin	kaempferol-3-O-rutinoside	*G. hirsutum*		[13, 16, 17]
Rhamnetin	7-O-methylquercetin	Flower petals from *G. hirsutum*		[18]
Tamarixetin	4'-O-methyl quercetin	Flower petals from *G. hirsutum*		[18]
Kaempferide	4'-O-methyl kaempferol	Flower petals from *G. hirsutum*		[18]
Quercetin	3, 5, 7, 3′, 4′-pentahydroxyflavone	Flower petals from *G. hirsutum*		[18]
Quercetin-3′-O-β-D-glucoside	quercetin-3′-O-β-D-glucoside	Flower petals from *G. hirsutum*		[19]
Quercetin-3-O-β-D-glucoside	quercetin-3-O-β-D-glucoside	Flower petals from *G. hirsutum*		[19]
Quercetin-7-O-β-D-glucoside	quercetin-7-O-β-D-glucoside	Flower petals from *G. hirsutum*		[19]
Isotrifoliin	quercetin 3-O-glucosside	Leaves from *G.* spp.		[14]
Spiraeoside	quercetin 4′-O-glucoside	Leaves from *G.* spp.		[14]
Hyperoside	quercetin-3-O-galactoside	Flower petals from *G. hirsutum*		[19]
Kaempferol	3, 4′, 5, 7-tetrahydroxyflavone	Flower petals from *G. hirsutum*		[13, 19]
Astragalin	kaempferol-3-O-glucoside	Flower petals and leaves from *G. hirsutum*		[14, 19, 20]
Kaempferol-3-O-β-D-(6″-O-p-coumaryl) glycoside	kaempferol-3-O-β-D-(6″-O-p-coumaryl) glycoside	Flower petals from *G. hirsutum*		[14, 19]
Rutin	quercetin-3-rhamnoglucoside	Seeds and flower petals from *G. hirsutum*		[14, 19, 21]
Isoquercitrin	quercetin-3-β-D-glucofuranoside	Flowers, leaves, & seeds from *G. hirsutum* type 108-F		[16, 21]
Butin	7, 3′, 4′-trihydroxyflavanone	Leaves from *G. spp.*		[14, 16, 21]
Hirsutrin	quercetin-3-β-D-glucopyranoside	Flowers & leaves from *G. hirsutum* type 108-F		[16, 21]
Hesperindin	3,5,7-trihydroxyflavanone-7-rhamnoglucoside			[22]

Name	Structure	Plant/part	Function	References
Hybridin	Quercetin-3-[1 ➔ 3-(β-D-xylopyranoside-β-D-glucopyranoside-β-D-galactofuranoside)]	Leaves of cotton plant of type 108-F and a hybrid of cotton plant with *Hibiscus*		[23]
Anthocyanins	quercetin-3-[1 ➔ 3-(β-D-xylopyranoside-β-D-glucopyranoside-β-D-galactofuranoside)]	Leaf	Reddening	[24]
Cyanidin	3, 3′, 4′, 5, 7-pentahydroxyflavylium	Leaves of *G. spp.*		[14, 25]
Cyanidin-3-β-glucoside	2-(3,4-dihydroxyphenyl)-3-(β-D-glucopyranosyloxy)-5,7-dihydroxy-1-benzopyrylium	Leaf	Resistance to Tobacco budworm	[26]
Cyanin	cyanidin-3,5-O-diglucoside	Leaves of *G.* spp.		[14]
Procyanidin B2	[(2*R*,3*R*,4*R*)-flavan-3,3′,4′,5,7-pentol]-(4 → 8)-[(2*R*,3*R*)-flavan-3,3′,4′,5,7-pentol]	Leaves of *G.* spp.		[14]
Delphinidin	3,3′,4′,5,5′,7-hexahydroxyflavylium			[14, 25]
Mirtillion	delphinidin 3-O-glucoside	Leaves of *G. spp.*		[14]
Gossypetin-8-O-rhamnoside	gossypetin-8-O-rhamnoside	*G. arboretum* tissues	Resistance to Tobacco budworm	[27]
Gossypetin-8-O-glucoside	gossypetin-8-O-glucoside	*G. arboretum* tissues	Resistance to Tobacco budworm	[27]
Catechin	3, 5, 7, 3′, 4′-pentahydroxyflavan	Stem steles		[28]
Gallocatchin	3, 5, 7, 3′, 4′, 5′-hexahydroxyflavan	Stem steles and leaves from *G.* spp.		[14, 28]
Gallocatchin-gallocatchin		Leaves from *G.* spp.		[14]
Tamarixetin	4’-O-methyl-quercetin	Flower petals from *G. hirsutum*		[18]
Kaemferide	4’-O-methyl-kaemferol	Flower petals from *G. hirsutum*		[18]
Dimethylated quercetin (Rhamnazin)	3′,7-di-O-methyl quercetin	Flower petals from *G. hirsutum*		[18]
Ombuin	4′,7-dimethylquercetin	Flower petals from *G. hirsutum*		[18]
Condensed proanthocyanidins		Vegetative tissue in both *G. hirsutum* and *G. barbadense*		[28]

Name	Structure	Plant/part	Function	References
Quercetin-3-O-neohesperidoside	3-[(2*S*,3*R*,4*S*,5*S*,6*R*)-4,5-dihydroxy-6-(hydroxymethyl)-3-[(2*S*,3*R*,4*R*,5*R*,6*S*)-3,4,5-trihydroxy-6-methyloxan-2-yl]oxyoxan-2-yl]oxy-2-(3,4-dihydroxyphenyl)-5,7-dihydroxychromen-4-one			[29]
Quercemeritrin	quercetin-7-O-β-D-pyronoside	Flowers from *G. hirsutum*		[30]
Tamarixetin	4'-O-methyl quercetin	*G.* sp.		[18]
Tamarixicitrin-7-glucoside	4'-O-methyl quercetin-7-O-glucoside	Flowers from *G.* sp.		[31]
Quercetin −3′-glucoside	quercetin −3′-glucopyranoside	Flowers from *G. hirsutum*		[16]
Spiraeoside	quercetin-4'-O-glucoside	Seeds from *G. hirsutum*		[16]
Quercetin-3-sophoroside	quercetin-3-O-[O-β-D-glucosyl-(1 → 2)- β-D-glucoside]	Flowers from *G. barbadense*		[31]
Hyperoside	quercetin-3-O-[O-β-D-glucosyl-(1 → 2) β-D-glucoside]	Flowers from *G. barbadense*		[31]
Naringenin	5,7,4′-trihydroxyflavone	Hypocotyls, ovules, fiber, roots, cotyledons, and leaves		[13, 32, 33]
Eriodictyol	(2S)-2-(3,4-dihydroxyphenyl)-5,7-dihydroxy-4-chromanone	Ovules, fiber, roots, cotyledons, and leaves from *G. hirsutum*		[33]
Herbacetin	3,4′,5,7,8-pentahydroxyflavone			
Herbacitrin	herbacetin 7-O-glucoside	Flowers from *G. arboreum* and *G. herbaceum*		[34]
Sexangularetin 3-glucoside 7-rhamnoside	8-O-methylherbacetin 3-O-glucoside 7-rhamnoside	Buds from *G. hirsutum*		[35]
Gossypetin	3,3′,4′,5,7,8-hexahydroxyflavone			
Gossypin	gossypetin 8-O-glucoside	Flowers from *G. arboretum* and *G. herbaceum*		[36, 37]
Gossypitrin	gossypetin 7-O-glucoside	Flowers from *G. arboretum* and *G. herbaceum*		[38, 39]
Gossypetin 8-O-rhamnoside	gossypetin 8-O-rhamnoside	Flowers from *G. arboretum*		[40]
Ilicicyanin	cyanidin 3-O-xylosylglucoside			[13]
Chrysanthemin	cynidin −3-O-β-D-glucoside			[41]

Name	Structure	Plant/part	Function	References
Gossypicyanin	cyanidin 3-O-β-D-xylosyl-β-glucoside			[13]
Leucocyanidin	(2R,3S,4S)-2-(3,4-dihydroxyphenyl)-3,4-dihydro-2H-chromene-3,4,5,7-tetrol	Anther, boll valves, stem bark, root bark		[42]
Leucodelphinidin	(2R,3S,4S)-2-(3,4,5-trihydroxyphenyl)-3,4-dihydro-2H-chromene-3,4,5,7-tetrol	Anther, boll valves, stem bark, root bark		[42]
Genistein	(2*S*)-4′,5,7-trihydroxyflavan-4-one	Hypocotyls from *G. hirsutum*		[32, 43]
Catechin	(2R,3S)-2-(3,4-dihydroxyphenyl)-3,4-dihydro-2H-chromene-3,5,7-triol	Leaves, hypocotyls, stem, cali, anthers, boll valves, stem bark, root bark, and cotton oil cake		[28, 32, 44–46]
(−)-Epicatechin	(−)-cis-3,3^{1},4^{1},5,7-pentahydroxyflavane	Anthers, boll valves, stem bark, root bark from *G. hirsutum*		[42]
(−)-Epigallocatechin	(−)-*cis*-3,3^{1},4^{1},5,5^{1},7-hexahydroxyflavane	Anthers, boll valves, stem bark, and root bark from *G. hirsutum*		[42]
Gallocatechin	(2S,3R)-2-(3,4,5-trihydroxyphenyl)-3,4-dihydro-1(2H)-benzopyran-3,5,7-triol	Hypocotyls/ stem steles, cali, anthers, boll valves, stem bark, and root bark from *G. hirsutum*		[28, 42]
Aromadendrin	(2R,3R)-3,5,7-trihydroxy-2-(4-hydroxyphenyl)-2,3-dihydrochromen-4-one	Ovules, fibers, roots, cotyledons, and leaves		[33]
Taxifolin	(2R,3R)-2-(3,4-dihydroxyphenyl)-3,5,7-trihydroxy-2,3-dihydrochromen-4-one	Ovules, fibers, roots, cotyledons, and leaves		[33]
Luteolin	3′,4′,5,7-Tetrahydroxyflavone			
Luteolin 8-C-hexosyl-O-hexoside	Luteolin 8-C-hexosyl-O-hexoside	Leaves from *G.* spp.		[14]
Pelargonidin	2-(4-hydroxyphenyl) chromenylium-3,5,7-triol			[14]
Pelargonin	3,5-*O*-diglucoside of pelargonidin			
Pelargonidin 3-O-beta-D-glucoside	pelargonidin 3-O-beta-D-glucoside	Leaves from *G.* spp.		[14]
Eriodictyol O-malonylhexoside	Luteolin 7-O-(6-O-malonyl-beta-D-glucoside)	Leaves from *G.* spp.		[14]
Peonidin	3,4′,5,7-tetrahydroxy-3′-methoxyflavylium			

Name	Structure	Plant/part	Function	References
Peonidin O-hexoside		Leaves from *G.* spp.		[14]
Naringenin	(2*S*)-4′,5,7-trihydroxyflavan-4-one	Leaves from *G.* spp.		[14]
Prunin	naringenin 7-O-glucoside	Leaves from *G.* spp.		[14]
Kuromanin	cyanidin 3-O-glucoside	Leaves from *G.* spp.		[14]

Table 1.
List various flavonoids reported in cotton.

2.1.2 Tannins

Tannins are polyphenolic substances that have molecular weights ranging from 500 to 3000 [47]. They can cause proteins to precipitate. Tannins possess numerous phenolic hydroxy groups that enable them to establish multiple hydrogen bonds with proteins and other large molecules with -NH-, $-NH_2$, or -OH groups [47]. This interaction results in the formation of complexes that can dissociate under normal pH conditions and are resistant to enzymatic degradation. The term “tannins” initially emerged from the leather industry, referring to substances capable of converting animal hides into leather [48]. Tannins can be categorized into two groups based on their biological origin. Hydrolyzable tannins are complex polyphenols that can be broken down through hydrolysis to yield gallic or ellagic acid and a sugar unit, typically glucose [47].

On the other hand, nonhydrolyzable or condensed tannins are formed by the polymerization of hydroxy flavans, resulting in the formation of dimers, trimers, or oligomers. Certain condensed tannins can become insoluble during extraction due to strong interactions with proteins, nucleic acids, or polysaccharides, leading to precipitate formation. Condensed tannins are also procyanidins since they produce anthocyanidins upon acid heating [48].

Together with phenolic acids [49, 50], gossypol [51–54], flavonoids [27], and tannins [47] are also listed in the groups of compounds that have been investigated and utilized in host resistance-breeding programs in cotton.

Condensed tannins have been documented as host plant resistance mechanisms for cotton bollworm (*Helicoverpa zea* Boddie) and tobacco budworm (*Heliothis virescens* F.). As a cotton resistance mechanism, condensed tannins were also found to antagonize the effects of the B.t. toxin [55]. The resistance imparted by condensed tannins deterred feeding by the larvae, which ultimately reduced the amount of B.t. toxin ingested by the insect; as a result, larvae feeding on high-tannin, B.t.-toxin-containing diets had lower mortality rates than insects feeding on diets containing condensed tannins or B.t. toxin alone. Condensed tannins could have also directly reduced the availability and activity of the B.t. toxin protein since tannins can precipitate proteins. Hence, the report proposed that insect management strategies utilizing B.t. toxin, whether as a topical insecticide or in genetically modified cotton crops, may not be suitable when combined with high-tannin plants.

Therefore, the report suggests that insect control tactics employing the B.t. toxin, either as a topically applied microbial insecticide or in transgenic cotton plants, may not be compatible with plants with high tannin concentrations.

Ref. [25] reported that condensed tannins in cotton leaves were up to 20% of dry weight. Their concentrations in individual leaves increased successively until about the tenth leaf. The upper leaves maintained the concentrations until early fall. The tannins resisted spider mites and could be hydrolyzed into digests consisting of one cyanidin and four delphinidin.

Tannins contribute to disease resistance. Resistant young leaves of cotton contain condensed tannins in higher concentration than do older susceptible leaves [56].

2.1.3 Lignins

Besides the above bioactive secondary metabolites, some structural metabolites such as lignins and cellulosic also belong to secondary metabolites [57], constituting a significant part of the biomass. For example, the cotton plant stalk comprises about 68% holocellulose and 26% lignin. These two classes of secondary metabolites also combine to be called lignocellulosic [1]. Lignins have protective functions, such as preventing lodging, etc.

2.1.4 Terpenoids

Terpenoids, or isoprenoids, are the largest group of plant secondary metabolites derived from the five-carbon compound isoprene, and its derivatives called terpenes. Terpenoids contain additional functional groups, usually oxygen, such as terpene aldehyde. Compared to flavonoids, cotton terpenoids do not have so many compounds, but the naked eye can see their existence as pigment glands. Cotton plants produce a variety of monoterpene and sesquiterpenoids. Monoterpenes include α-pinene, β -pinene, limonene, β-ocimene, β-myrcene, α-terpenene, γ-terpenene, etc. They are volatile oils, often as semiochemicals, attracting or deterring insects or parasitoids [34, 58]. Cotton plants produce both volatile and nonvolatile sesquiterpenoids. The former includes α-humulene, α-copaene, β-carophyllene, β-carophyllene oxide, α-aromadendrene, α-selinene, β-selinene, γ-bisabolene, β-bisabolol, δ-cadinene, 12-hydroxy-β-caryophyllene, gossonorol, spathulenol, 12-hydroxy-β-caryophyllene acetate, 12-hydroxy-β-caryophyllene oxide acetate, and guaia-1(10), 11-diene, etc. The latter includes hemigossypol, hemigossypolone, desoxyhemigossypol, gossypol, and heliocides H_1, H_2, H_3, and H_4 [59], as well as 6-methoxy and 6, 6′-dimethoxy gossypol derivatives [60]. They are nonvolatile. Almost all sesquiterpenoids and part of monoterpenes (the remaining are released to the air) are stored in the pigment glands, which appear oval and spherical and as black, orange, yellow-brown (yellowish-brown), green (red-brown), or purple depending on the species in all the tissues of cotton plants except pollen and seed coat [61]. Some monoterpenes (β-ocimene and myrcene) and sesquiterpenoids (hemigossypolone) form heliocides (C-25) H_1, H_2, H_3, and H_4) by direls alder reaction in the glands. Different genotypes impacted the distribution of the pigment glands [62, 63]. **Figure 2** shows the structures of some common terpenoids in cotton plants.

Gossypol is a dimeric sesquiterpenoid stored in the pigment glands of cotton plants. Gossypol was first extracted from cotton seeds by Marchlewski in 1899 and characterized by Adams et al. in 1938 [64, 65]. It constitutes 20–40% of the pigment gland weight and accounts for 0.4–1.7% of the whole cotton seed kernel. From the gossypol history, gossypol was first discovered from cotton seeds (cotyledons) because the toxicity of gossypol to humans and animals blocked the efficient use of cottonseed oil and protein. However, [66] found the importance of cotton roots in

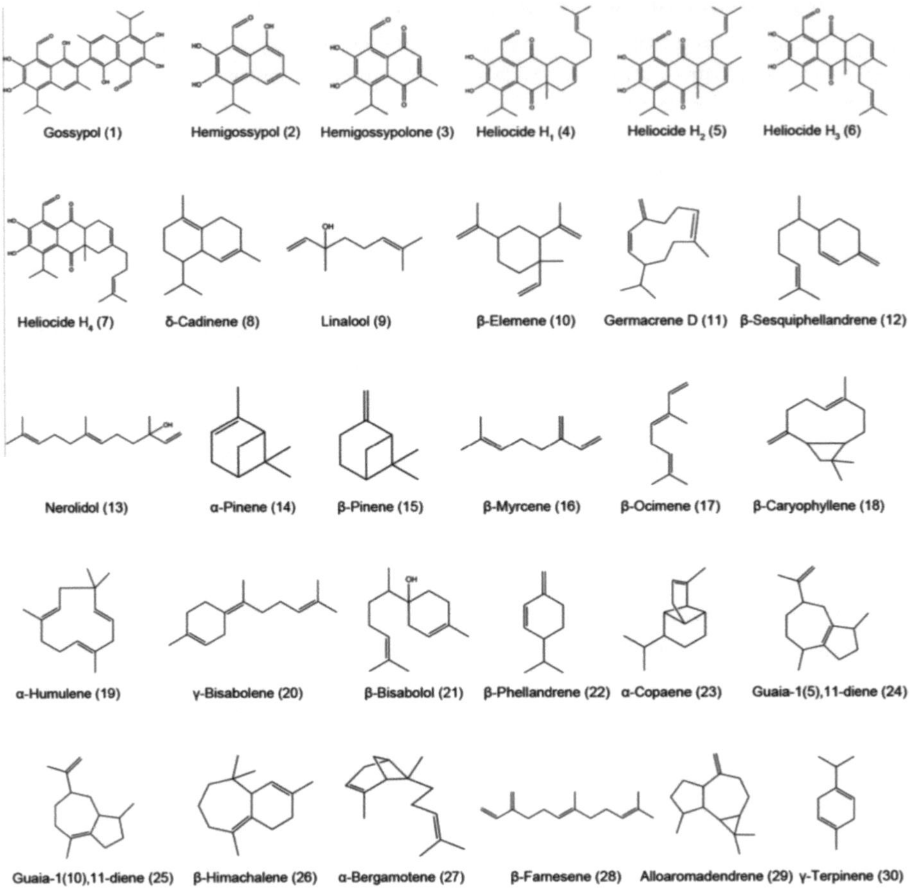

Figure 2.
Structures of some common terpenoids in cotton plants.

gossypol production in cotton plants. Recent progress [61] showed that cotyledons and roots are primary sources of gossypol in cotton plants. These two plant parts contain higher gossypol than other parts. The former is responsible to provide gossypol during germination, after which the developing roots become the primary source for gossypol production in cotton plants.

The defensive role of cotton pigment glands and gossypol was first recognized by Cook [67] in 1906. It turned out that gossypol and other terpenoids in the cotton pigment glands are associated closely with resistance mechanism to insects and pathogens [68, 69], including monoterpenes, sesquiterpenoids, and heliocides.

Despite that the pigment glands are distributed almost throughout the whole plant, the gland content compositions are different from different parts. The glands of cottonseeds contain mostly gossypol with traces of deoxyhemigossypol, and the glands of cotton leaves contain hemigossypolone, a gossypol derivative, and heliocides [70]. On fresh weight basis, cotton seed (cotyledons) usually has similar gossypol content of around 1% in roots at the harvesting stage (Yue, unpublished data); on dry weight basis, cotton root usually has the highest gossypol content [66] among different plant parts, which is consistent with the recent progresses that cotyledon (seeds) and roots are the source of gossypol for the whole plants [61, 71].

Name	Structure	Plant/part	Reference
Vanillic acid	3-methoxy-4-hydroxybenzoic acid	Seed	[72]
Synaptic acid	3, 5-dimethoxy-4-hydroxybenzoic acid	Seed	[72]
p-coumaric acid	4-hydroxy-cinnamic acid	Seed	[72]
Ferulic acid	3-methoxy-4-hydroxycinnamic acid	Seed	[72]
Syringic acid	3, 5-dimethoxy-4-hydroxycinnamic	Seed	[72]
p-hydroxybenzoic acid		Seed	[72]
p-hydroxybenzoic acid	4-hydroxybenzoic acid	Seed	[72]
o-coumaric acid	2-hydroxycinnamic acid	Seed	[72]
gentisic acid	2, 5-dihydroxybenzoic	Seed	[72]
caffeic acid	3, 4-dihydroxycinnamic	Seed	[72]
protocatechuic acid	3, 4-dihydroxybenzoic	Seed	[72]
Chlorogenic acid	3-*O*-Caffeoylquinic acid	Seed	[72]
Lignin	phenolics	Stem	[73]

Table 2.
Phenolic acids in cotton.

2.1.5 Phenolic acids

Phenolic acids in cotton include benzoic acid, cinnamic acid, and their derivatives (**Table 2**). They are water-soluble. They include vanillic acid, benzoic acid, ferulic acid, sinapic acid, cinnamic acid, syringic acid, p-coumaric acid, gentisic acid, caffeic acid, chlorogenic acid, gentisic acid, p-hydroxybenzoic acid, p-hydroxybenzoic acid, 3, 4-dihydroxybenzoic (protocatechic) acid, etc. Their functions in cotton plants were thought to be related to resistance mechanism to pests and pathogens [50]. They also played a role in cotton allelopathy [74].

2.1.6 Fatty acids

Cotton fatty acids are mainly in the seeds. Cottonseed oil has a fatty acid profile that is composed of 52.89% linoleic acid, 25.39% palmitic acid, 16.35% oleic acids, together with small amounts of 2.33% stearic acid, 1% myristic acid, and 0.6% palmitoleic acid, as well as 0.17% linolenic acid [75]. Ref. [76] reported malvalic, sterculic, and dihydrosterculic acids in the hypocotyl of both glanded and glandless cotton seeds. Malvalic and sterculic acids represent 0.3% and 0.5% of the total fatty acids in the analyzed varieties.

2.1.7 Miscellaneous secondary compounds

Beyond these main classes of secondary metabolites, the following secondary compounds were also identified in the cotton plant (**Table 3**).

Name	Structure	Plant part	Reference
Glycine-betaine	(trimethylammoniumyl) acetate		[77]
Scopoletin	7-Hydroxy-6-methoxy-2H-chromen-2-one	Dried Bract of the Cotton Plant	[49]
Piceatannol	3′,4′,3,5-Tetrahydroxy-*trans*-stilbene Astringinin		[43]
Resveratrol	*trans*-3,5,4′-Trihydroxystilbene		[43]
Astragalin	3-(β-D-Glucopyranosyloxy)-4′,5,7-trihydroxyflavone		[43]
Pterostilbene	3,5-Dimethoxy-4′-hydroxy-*E*-stilbene		[43]

Table 3.
Additional secondary compounds identified in the cotton plant.

2.2 Secondary metabolite compositions of different cotton species

Compositions of secondary metabolites vary among cotton species. From 1-week-old roots of *Gossypium hirsutum* and *G. barbadense*, the disesquiterpenoid aldehydes, gossypol, 6-methoxygossypol, and 6, 6'-dimethoxygossypol, and the sesquiterpenoid aldehydes, hemigossypol and methoxyhemigossypol were isolated and identified. While the compounds 6-methoxygossypol and 6, 6′-dimethoxygossypol constituted 30% of the total terpenoid aldehydes in the seeds of the cultivar of *G. barbadense* but occurred only in trace quantities in those of *G. hirsutum* [60].

As mentioned above, the cotton genus contains around 50 species, and four are domesticated independently [2, 78]. These different cotton species have other secondary metabolite profiles, which are representative of biodiversity. For example, myrcene is a major monoterpene presented in *G. hirsutum* but not present in *G. barbadense*. In contrast, *G. barbadense* contains copaene, β-carophyllene oxide, and (−)-δ-cadinene at 14.3, 8.0, and 7.8% in leaf oil compared to only 0.7, 0.4, and 0.3 respectively in oils from *G. hirsutum*. The variations between the two species could potentially impact insect-host relationships [79].

Different biotypes in the same species also have different secondary metabolite compositions. Ref. [80] found high levels of intraspecific diversity in the terpene profiles of the plants. Two distinct chemotypes were identified: one chemotype contained higher levels of the monoterpenes γ-terpinene, limonene, α-thujene, α-terpinene, terpinolene, and p-cymene, while the other chemotype was distinguished by higher levels of α- and β-pinene. The distribution of chemotypes followed a geographic gradient from west to east with an increasing frequency of the former chemotype. Such differences in secondary metabolites are the foundation for a cotton breeding program for a high level of one or several secondary metabolites.

2.3 Inductive vs. constitutive secondary metabolites

A secondary metabolite can be constitutive or inducive. A constitutive secondary metabolite means the plant produces the secondary metabolite as a routine without environmental influence. An inductive secondary metabolite is produced or increased under one or more environmental factors, such as pest infestation or abiotic stresses. Most secondary metabolites are both constitutive and inductive. For example, cotton flavonoids and gossypol are constitutive as they are routinely produced, and they are inductive as their production increases when there are abiotic stresses, such as salinity stress [81]. In addition, increased terpenoid accumulation in cotton foliage was observed as a general wound response upon mechanical damage, *Spodoptera littoralis* caterpillar infestation, and jasmonic acid treatment [82] generally increased secondary metabolites production with E-β-ocimene, heliocide H_1 and H_4, showing the highest increases. In response to aphid attack, cotton plants synthesized and released callose outside the cell plasma membrane [83].

2.4 Cotton secondary metabolites and abiotic and biotic stresses

2.4.1 Abiotic stresses

Cotton production can be affected by various abiotic stresses, which include drought, salinity, nutrient deficiency, chemical burns, and ultraviolet radiation stress. They all turned out to have interactions with secondary metabolites. Abiotic stresses can cause secondary metabolite content and pest resistance change, influencing fiber and seed quality.

Ref. [84] reported total phenol and tannin content under salinity stress increased significantly after third water stress exposure. Evidence also showed that secondary metabolite increase resulted from abiotic stresses, increased pest resistance, and led to higher fiber quality. Such cotton traits can be applied in cotton production directly.

Nutrient deficiency is another kind of abiotic stress, and Ref. [85] showed that K deficiency reduced the antioxidant capacity of cotton seedlings and resulted in a metabolic disorder characterized by elevated levels of primary metabolites and reduced production of secondary metabolites. The results were from an analysis of xylem sap. How much degree the xylem sap represents the defense response of cotton plants to K deficiency is of concern. We compared root gossypol concentrations of cotton seedlings from irrigated under tap water and Hoagland solution for three weeks after emergence. The root gossypol concentrations of tap water irrigated seedlings (subjected to nutrient deficiency) were higher than the Hoagland solution irrigated seedling (Yue, unpublished data).

Cotton plants suffer from ultraviolet radiation damage, but many cotton secondary metabolites can protect against ultraviolet radiation. This partly originates from the fluorescence properties of some secondary metabolites that can convert ultraviolet radiation to lower-energy light. It partly results from the solid antioxidative properties of cotton secondary metabolites [86].

2.4.2 Biotic stresses

Biotic stresses in cotton production include insects and other herbivores, pathogens, weeds, etc. As discussed in 2.3, the secondary metabolite production was a defense response to the insects and salinity stresses; pathogen inoculation/ infection also induced secondary metabolite biosynthesis. Ref. [28] used dimethoxybenzaldehyde (DMB) histochemical reagent to reveal induced biosynthesis of flavanols (catechin, gallocatechin, and their condensed proanthocyanidins) as red color in the cotton stem section upon *Verticillium dahliae* inoculation.

The *Verticillium* wilt-resistant cultivar showed much more intense flavanol synthesis than the susceptible cultivar.

Among the different stresses for cotton plant growth, development, and reproduction such as insects, pathogens, and abiotic stresses (drought, salinity, and nutrient deficiency, etc.), only weed infestation has not been reported to induce plant defense or immune response or change of secondary metabolites concentrations, which is the basis for potential utilization. Phenolic acids have been explored for their weed-suppressive effects in cotton [74]. Flavonoids, however, have not been examined. Cotton has a rich array of flavonoids, some of which, such as quercetin, have been reported to be allelopathic [87].

2.5 Glandless cotton

Since McMichael bred a glandless cotton cultivar in 1959 [88], which was devoid of toxic gossypol, the utilization of cotton secondary metabolites has another aspect: reduction of secondary metabolites so that cotton seeds can be less toxic and more valuable. Recently, Ref. [89] reported the use of RNAi to selectively silence the δ-cadinene synthase gene to reduce gossypol levels in cotton seed by 97% while not affecting levels of gossypol and other terpenoids in the rest of the plants.

3. Utilization of secondary metabolites in cotton production

3.1 Secondary metabolites being extracted and applied directly in cotton production

Cotton secondary metabolites have been recognized as natural pesticides [79]. The simplest way to utilize them is to extract them and apply the extract directly in cotton production.

Ref. [90] extracted secondary metabolites by water from different plant parts of *G. hirsutum* plant and used them to suppress weed. Among the various water extracts of the plant parts, leaf extract was found to impart maximum inhibitory effect on the germination indices of the wheat (*T. aestivum*) followed by the stem extracts. The concentrated extract led to wheat germination percentage and seedling vigor index reduction of 46% and 62%, respectively. The research did not reach into the active ingredients of the cotton leaf extract. As discussed earlier, cotton leaf is rich in secondary metabolites such as condensed tannins (up to 20% dry weight), terpenoids, non-tannin flavonoids, and phenolic acids. As terpenoids are not soluble in water, they might not be the main active ingredients of the extract except the extract contained a significant number of solid particles (our HPLC analysis showed leaf water leachate contained less than 1 mg/L gossypol). Hence, the most probable active ingredients are non-tannin flavonoids and phenolic acids. Phenolic acids have been explored for cotton allelopathy [74]. Flavonoids, however, have not been investigated for cotton allelopathy. As listed in **Table 1**, cotton contains a rich array of flavonoids, some of which have been reported to be allelopathic. For example, Ref. [87] reported quercetin from *Fagopyrum esculentum* roots to inhibit the radicle growth of *Phelipanche ramose*. The data also suggest that two ortho-free hydroxy groups of the C (B?) ring could be essential to impart allelopathic activity. In contrast, the carbon skeleton of the B ring and substituents of both the A and B rings are unnecessary. According to this rule, lots of flavonoids in **Table 1** are allelopathic such as Gossypetin, Gossypin,

Gossypitrin, Gossypetin 8-O-rhamnoside, Catechin, Luteolin, Luteolin 8-C-hexosyl--O-hexoside, Hyperoside, Quercetin-3-sophoroside, Quercemeritrin, Gallocatchin, Delphinidin, Cyanidin, Hirsutrin, Butin, Isoquercitrin Hyperoside, Isotrifoliin, Rhamnetin, etc. Hence, allelopathy spread by cotton flavonoids is expected to be confirmed soon [91].

In this way of extraction and application, we can extract secondary metabolites from other allelopathic plants and apply them to cotton production. For example, Ref. [92] used water extracts from sorghum, sunflower, and brassica for preemergent application in the cotton field. It was observed that the best treatment produced dry biomass reduction by 40% of weeds *Trianthema portulacastrum* and *Cyperus rotundus* and an increase in seed cotton yield (12%). Similar extraction and application can be extended to different allelopathic plants. For example, we used sicklepod seed methanol extract (replaced with water after extraction) and applied it to soybean to prevent deer browsing [93].

3.2 Screening and breeding for high secondary metabolites cultivars

Since Cook [67] observed the better resistance to insect herbivores of Guatemalan cotton, numerous efforts have been made to increase U.S. cotton resistance to various biotic and abiotic stresses in cotton production [6]. So far, secondary metabolites were reported to contribute cotton resistance include condensed tannins [47, 89], terpenoids/gossypol [51–54, 63, 94], phenolic acids [49, 50, 74], and bacterial endotoxins [95, 96]. All these secondary metabolites have been screened and bred for higher secondary metabolite levels. Their common basis is that selection among cultivars or species for higher secondary metabolite levels is an attainable goal [63, 97].

3.3 Cotton secondary metabolite regulators or elicitors

As discussed before, cotton secondary metabolites (gossypol and flavonoids) were shown to increase under salinity stress [81]. The salinity stress was imposed by salt water as irrigation; foliar application of the salt water also increased root gossypol concentrations in our lab. The saltwater can be regarded as a secondary metabolite regulator or elicitor to increase gossypol and flavonoid production for defense. In addition, Ref. [43] showed that oligosaccharides filtrates of *Fusarium oxysporum* f. sp. *vasinfectum* could stimulate phenolic compounds production in cotton plants. Furthermore, some herbicides [98, 99] were reported to cause an increase in gossypol concentration in cotton plants. Since the introduction of GMO cotton in the mid-1990s, herbicide application has become a routine and indispensable practice in cotton production. Tank-mix of herbicide with secondary metabolites regulators or elicitors, such as the salt, as mentioned earlier solution or pretreatment of the cotton seeds by the herbicides as mentioned above can increase the gossypol concentration of the seedling roots, a potential practice to prevent nematode damage to cotton plants.

3.4 Metabolic engineering method to modify cotton

Contrary to the exogenous metabolite regulator, metabolic engineering optimizes genetic and regulatory processes within cells to increase the cell's production of a specific substance. Compared to general metabolic engineering, where the target substance is commercial, we would increase specific substances (secondary metabolites) in situ for cotton plant defense or profit.

Cotton is a multibillion-dollar industry where engineering is often employed to improve profitability. In a study conducted by John and Keller in 1996 [100], they utilized particle bombardment to introduce phbB and phbC genes into cotton. This genetic modification led to the development of cotton plants with fibers containing poly-(3R)-hydroxybutanoate (PHB), thereby enhancing the insulating properties of the fibers. Genetic mapulations of production of trienoic fatty acids in cotton seeds have improved low-temperature seed germination, plant photosynthesis, and fiber quality [101].

Most recently, Ref. [102] used a selective gene editing method to remove the toxic enantiomer (−)-gossypol while the (+)-gossypol and other structurally related terpenoids remained, keeping their resistance to insect herbivores and pathogens. The modified genes (including the current widely accepted GMO cotton genes) may flow into wild species in nature, which is a concern.

3.5 Cotton secondary metabolites as a potential stress assessment tool for field crop management

Although the general concept of plant secondary metabolites and their ecological functions became evident around the mid-twentieth century, gossypol was first extracted from cotton seeds in 1899 [103], and cotton pigment glands were correctly interpreted in 1906 [67]. Hence, we have accumulated over 124 years of knowledge of cotton secondary metabolites. The knowledge of cotton secondary metabolites has benefitted cotton breeding and human and animal health regarding contraception, byssinosis, and utilization of cotton seeds. Now, we have known draught [84], salinity [81], chilling [104], insect [86], pathogen [59], nutrient deficiency (Yue, unpublished data), herbicides [98, 99], and even mechanical damage [82] of the cotton plants can induce production of cotton terpenoids in cotton plants, especially root gossypol concentration to increase. In addition, the root system is the main organ to synthesize gossypol [61]. The root system also increased its gossypol concentration while supplying gossypol to aboveground parts except cotyledons. Hence, cotton root gossypol concentration is sensitive to all the above factors and is a cumulative index of cotton plant stress and defense status. In other words, cotton root gossypol concentrations are the common currency in cotton plant stress and defense response.

Furthermore, our ongoing projects showed that adjacent plants can increase cotton root gossypol concentration compared to isolated cotton plants without adjacent plants [105]. Nutrient deficiency stress increased cotton root gossypol concentration comparing to cotton plants fertilized by Hoagland solution (Yue, unpublished data). By analyzing all the previous data on cotton root gossypol concentrations, it is found cotton roots have the highest gossypol concentration on a dry matter basis (6.3% in Ref. [66]) among different cotton plant parts. Cotton roots have a similar gossypol concentration as seeds at the harvesting stage on a fresh weight basis, around 1% (Yue, unpublished data).

Cotton root gossypol concentration analysis is potentially a stress assessment tool for field crop management, just like body temperature measurement in medical examinations. The basis is that we set up a series of standard values of cotton root gossypol concentrations at different ages or sizes without those stresses.

It is time to push our understanding of the cotton secondary metabolites into productivity. Direct application of cotton secondary metabolites in crop management will convert the traditional knowledge accumulation pattern of cotton secondary metabolites from funding agency publication to application data production, making us acquire knowledge of cotton secondary metabolites faster. A specific farmer might not be able to analyze these secondary metabolites, but our research and extension centers should be able to do it.

3.6 Employment or avoidance of effects of secondary metabolites in cotton stubble on soil health and cotton crop next season as another field of application of cotton secondary metabolites in cotton production

As a cumulative stress index, cotton root gossypol concentration increases with age and size. At harvesting stage, cotton root gossypol concentration is high and around 1% (Yue unpublished data) fresh weight, each cotton stubble is estimated as 10 grams, and cotton plant density is estimated as 40,000 plants/acre [106]. The phytoalexin gossypol dose is estimated as 4000 g/acre, this dose is significantly heavier comparing to other herbicides and insecticides. For example, glyphosate burndown dose is 1 lb./acre [107]. Despite constrained in the dead cotton stubbles, the root gossypol and its glands are expected to have significant impact on soil health and possibly cotton crop next season. Employment of their effects on soil health and cotton plants next season and avoidance of the effects are another field of application of secondary metabolites in cotton production.

4. Conclusions

Cotton plants are rich in secondary metabolites, which mainly include terpenoids, flavonoids, and phenolic acids. Terpenoids are not soluble in water, stored in pigment glands, including monoterpenes, sesquiterpenoids, and heliocides. They can deter insects and pathogens. One representative is gossypol, whose main sources are cotyledons and living roots. Cotton root gossypol concentration is a cumulative index for cotton plant stresses and defense response, and potentially used as a stress assessment tool for cotton crop management. It also increases with cotton plant age and size, reaching around 1% root fresh weight at harvesting stage and representing a heavier dose than other herbicides and insecticides. The impact of the high-dose gossypol in cotton roots after harvest on soil health and cotton plants next season is expected to be confirmed in the near future. Flavonoids used to be thought only as pigments but recent literature showed they can also deter insects and pathogens. A key difference of flavonoids from terpenoids is that they are soluble in water and stored in vacuoles. The allelopathic properties of flavonoids are expected to be confirmed in the near future. Tannins are a special group of flavonoids as they can bond to and precipitate protein and are also involved in deterring insects and pathogens. Phenolic acids are often precursors in flavonoids synthesis and soluble in water. Literature shows they are active to deter insects and pathogen and also allelopathic to suppress weeds.

Traditionally, all the three classes of cotton secondary metabolites terpenoids (gossypol), flavonoids, including tannins and phenolic acids, are integrated in cotton screening and breeding programs. Recent progresses have expanded their application in cotton production; direct extraction of the secondary metabolites and application; use of chemical regulators or elicitors of cotton secondary metabolites to change their concentrations *in situ* to achieve the functions of the target secondary metabolites, and change of the genetic bases of the secondary metabolites to produce or not to produce one or a set of secondary metabolites in crop plants *in situ* to achieve some benefits, which include reduction of the toxic gossypol in cotton seeds to increase their value in utilizing their oil and protein.

Author details

Ziming Yue*, Te-Ming Tseng, K. Raja Reddy, Natraj Krishnan and Shien Lu
Mississippi State University, Mississippi State, USA

*Address all correspondence to: ziming.yue@gmail.com

References

[1] Velmourougane K, Blaise D, Savitha S, Waghmare VN. Chapter 34 – Valorization of cotton wastes for agricultural and industrial applications: Present status and future prospects. In: Bhat R, editor. Valorization of Agri-Food Wastes and by-Products. Cambridge: Academic Press; 2021. pp. 665-692

[2] Grover CE, Grupp KK, Wanzek RJ, et al. Assessing the monophyly of polyploid Gossypium species. Plant System Evolution. 2012;**298**:1177-1183. DOI: 10.1007/s00606-012-0615-7

[3] Gallagher JP, Grover CE, Rex K, Moran M, Wendel JF. A new species of cotton from wake atoll, *Gossypium stephensii* (Malvaceae). Systematic Botany. 2017;**42**:1548-2324. DOI: 10.1600/036364417X694593

[4] Bell AA. Physiology of secondary products. In: Mauney JR, Stewart JMCD, editors. Cotton Physiology. Memphis: The Cotton Foundation; 1986. pp. 597-621

[5] Stipanovic RD, Williams HJ, Bell AA. Secondary products. In: Stewart JM, Oosterhuis DM, Heitholt JJ, Mauney JR, editors. Physiology of Cotton. Dordrecht: Springer; 2010. pp. 342-352. DOI: 10.1007/978-90-481-3195-2_30

[6] Nix A, Paull C, Colgrave M. Flavonoid profile of the cotton plant, *Gossypium hirsutum*: A review. Plants (Basel). 2017;**6**(4):43. DOI: 10.3390/plants6040043

[7] Did you know? 65, 66. In: Learning Corner. Available from: https://icac.org/LearningCorner/LearningCorner?CategoryId=1&MenuId=15 [Accessed: November 28, 2023]

[8] Vulchi R, Bagavathiannan M, Nolte SA. History of herbicide-resistant traits in cotton in the U.S. and the importance of integrated weed management for technology stewardship. Plants (Basel). 2022;**11**(9): 1189. DOI: 10.3390/plants11091189

[9] Anjali SK, Korra T, Thakur R, Arutselvan R, Kashyap AS, Nehela Y, et al. Role of plant secondary metabolites in defense and transcriptional regulation in response to biotic stress. Plant Stress. 2023;**8**:100154

[10] Wink M. Plant secondary metabolism: Diversity, function and its evolution. Natural Product Communications. 2008;**3**:1205-1216

[11] Amitav BA. Chapter 5 – High-temperature stress and metabolism of secondary metabolites in plants. In: Bhattacharya A, editor. Effect of High Temperature on Crop Productivity and Metabolism of Macro Molecules. Academic Press; 2019. pp. 391-484. DOI: 10.101

[12] Croft KD. The chemistry and biological effects of flavonoids and phenolic acids. Annals of the New York Academy of Sciences Journal. 1998;**854**: 435-442

[13] Ismailov AI, Karimdzhanov AK, Islambekov SY, Rakhimkhanov ZB. Flavonoids of the cotton plant and plants close to it. Chemistry of Natural Compounds. 1994;**30**:1-14

[14] Long L, Zhao X-T, Feng Y-M, Fan Z-H, Zhao J-R, Wu J-F, et al. Profile of cotton flavonoids: Their composition and important roles in development and adaptation to adverse environments. Plant Physiology and Biochemistry. 2023;**201**:107866

[15] Parks CR. Floral pigmentation studies in genus *Gossypium*. 3.

Qualitative analysis of total flavonol content for taxonomic studies. American Journal of Botany. 1967;**54**:306-315

[16] Pakudina ZP, Sadykov AS. Isoastragalin—A flavonol glycoside from the flowers of *Gossypium hirsutum*. Chemistry of Natural Compounds. 1970; **6**:24-25

[17] Pratt C, Wender SH. Identification of Kaempferol-3-rhamnoglucoside and Quercetin-3-glucoglucoside in cottonseed. Journal of the American Oil Chemists' Society. 1961;**38**:403-405

[18] Struck RF, Kirk MC. Methylated flavonols in genus *Gossypium*. Journal of Agricultural and Food Chemistry. 1970; **18**:548-549

[19] Wu T, Abdulla R, Yang Y, Aisa HA. Flavonoids from *Gossypium hirsutum* flowers. Chemistry of Natural Compounds. 2008;**44**:370-371

[20] Zhang Y, Li W, Hu Y, Ding T, Zafar MM, Jia X, et al. Cotton flower metabolites inhibit SARS-CoV-2 main protease. FEBS Open Biology. 2022;**12** (10):1886-1895. DOI: 10.1002/2211-5463.13477. Epub 2022 Sep 17

[21] Pratt C, Wender SH. Identification of rutin and isoquercitrin in cottonseed. Journal of the American Oil Chemists' Society. 1959;**36**:392-394

[22] Shaver TN, Lukefahr MJ. Effect of flavonoid pigments and gossypol on growth and development of the bollworm, tobacco budworm, and pink bollworm. Journal of Economic Entomology. 1969;**62**:643-646. DOI: 10.1093/jee/62.3.643

[23] Makhsudova B, Pakudina ZP, Sadykov AS. The structure of hybridin. Chemistry of Natural Compounds. 1969; **5**:273

[24] Edreva A, Gurel A, Gesheva E, Hakerlerler H. Reddening of cotton (*Gossypium hirsutum* L.) leaves. Biologia Plantarum. 2002;**45**:303-306

[25] Lane HC, Schuster MF. Condensed tannins of cotton leaves. Phytochemistry. 1981;**20**:425-427

[26] Hedin PA, Jenkins JN, Collum DH, White WH, Parrott WL. Multiple factors in cotton contributing to resistance to the tobacco budworm, *Heliothis-Virescens* F. ACS Symposium Series. 1983;**208**:347-365. DOI: 10.1021/bk-1983-0208.ch020

[27] Hedin PA, Jenkins JN, Parrott WL. Evaluation of flavonoids in *Gossypium arboreum* (L.) cottons as potential source of resistance to tobacco budworm. Journal of Chemical Ecology. 1992;**18**: 105-114

[28] Mace ME, Bell AA, Stipanovic RD. Histochemistry and identification of flavanols in *verticillium* wilt-resistant and -susceptible cottons. Physiological Plant Pathology. 1978;**13**:143-149

[29] Blouin FA, Zarins ZM. High-perfomance liquid chromatography of *cottonseed* flavonoids. Journal of Chromatography A. 1988;**441**:443-447

[30] Attree GF, Perkin AG. The position of the sugar nucleus in the quercetin glucosides. Journal of Chemical Society. 1927;**120**:234-240

[31] Parks CR, Ezell WL, Williams DE, Dreyer DL. Symposium on biochemical sytematics, genetics and origin of cultivated plants. 7. Application of flavonoid distribution to taxonomic problems in genus *Gossypium*. Bulletin of the Torrey Botanical Club. 1975;**102**:350-361

[32] Kouakou TH, Waffo-Téguo P, Kouadio YJ, Valls J, Richard T, Decendit

A, et al. Phenolic compounds and somatic embryogenesis in cotton (*Gossypium hirsutum* L.). Plant Cell, Tissue and Organ Culture. 2007;**90**:25-29. DOI: 10.1007/s11240-007-9243-2

[33] Tan J, Tu L, Deng F, Hu H, Nie Y, Zhang X. A genetic and metabolic analysis revealed that cotton fiber cell development was retarded by flavonoid naringenin. Plant Physiology. 2013;**162** (1):86-95. DOI: 10.1104/pp.112.212142

[34] Röse US, Tumlinson JH. Volatiles released from cotton plants in response to *Helicoverpa zea* feeding damage on cotton flower buds. Planta. 2004;**218**: 824-832. DOI: 10.1007/s00425-003-1162-9

[35] Elliger CA. Sexangularetin 3-glucoside-7-rhamnoside from *Gossypium hirsutum*. Phytochemistry. 1984;**23**:1199-1201

[36] Rao KV, Seshadri TR. Constitution of gossypin-Part I. Proceedings Indian Academy of Sciences. 1946;**24A**:375-381

[37] Rao KV, Seshadri TR. Constitution of gossypin-Part II. Proceedings Indian Academy of Sciences. 1947;**25A**:397-403

[38] Subramanian SS, Nair AGR. Flavonoids of four Malvaceous plants. Phytochemistry. 1972;**11**(4):1518-1519

[39] Nair AGR, Subramanian SS, Swamy MN. Pigments of the flowers of Hibiscus surattensis. Current Science. 1962;**31**(9): 375-376

[40] Waage SK, Hedin PA. Biologically-active flavonoids from *Gossypium arboretum*. Phytochemistry. 1984;**23**: 2509-2511

[41] Rakhimkhanov ZB, Karimdzhanov AK, Ismailov AI, Sadykov AS. A study of the anthocyanins of the cotton plant. Chemistry of Natural Compounds. 1968; **4**:190-191

[42] Karimdzhanov AK, Kuznetsova NN, Dzhataev SA. Phenolic compounds of the plant *Gossypium hirsutum* and of callus tissue from its anthers. Chemistry of Natural Compounds. 1997;**33**:187-189

[43] Bla NARE, Hilaire KT, François KYK, Mongomaké K, Daouda K. Stimulation of phenolic compounds production in cotton (*Gossypium hirsutum* L.) by oligosaccharide filtrates of *fusarium oxysporum f. sp. Vasinfectum*. Chemical Science Review and Letters. 2015;**4**(15):788-797

[44] Rani PU, Pratyusha S. Defensive role of *Gossypium hirsutum* L. anti-oxidative enzymes and phenolic acids in response to *Spodoptera litura* F. Feeding. Journal of Asia-Pacific Entomology. 2013;**16**:131-136

[45] Kouakou TH, Kouadio YJ, Teguo PW, Valls J, Badoc A, Merillon JM, et al. Polyphenol levels in two cotton (*Gossypium hirsutum* L.) callus cultures. Acta Botanica Gallica. 2009;**156**:223-231

[46] Zhang HL, Nagatsu A, Okuyama H, Mizukami H, Sakakibara J. Sesquiterpene glycosides from cotton oil cake. Phytochemistry. 1998;**48**:665-668

[47] Legé KE. Tannins in cotton. In: Bajaj YPS, editor. Cotton. Biotechnology in Agriculture and Forestry. Vol. 42. Berlin, Heidelberg: Springer-Verlag; 1998. pp. 350-363. DOI: 10.1007/978-3-642-80373-4_23

[48] Swain T. Phenolics in the environment. In: Swain T, Harborne JB, Van Sumere CF, editors. Recent Advances in Phytochemistry,Vol 12. New York: Plenum Press; 1979. p. 624-637

[49] Wakelyn PJ, Stipanovic RD, Bell AA. Identification of scopoletin in dried bract of the cotton plant. Journal of Agricultural and Food Chernistry. 1974; **22**:567-568

[50] Lege KE, Cothren JT, Smith CW. Phenolic acid and condensed tannin concentrations of six cotton genotypes. Environmental and Experimental Botany. 1995;**35**(2):241-249

[51] Dilday RH, Shaver TN. Variability in flower-bud gossypol content and agronomic and fiber properties within the primitive race collection of cotton. Crop Science. 1980;**20**:91-95

[52] Jaroszewski JW, Strom-Hansen T, Hansen LL. Optical stability of gossypol. Chirality. 1992;**4**:216-221

[53] Hedin PA, Shepherd RL, Kappelman AJ Jr. Evaluation of cotton polyphenols as factors of resistance to root-knot nematode and *fusarium* wilt. Journal of Agricultural and Food Chemistry. 1984; **32**:635-638

[54] Shaver TN, Dilday RH, Wilson FD. Use of glandless breeding stocks to evaluate unknown *Heliothis* growth inhibitors (X-factors) in cotton. Crop Science. 1980;**20**:545-548

[55] Navon A, Hare JD, Federici BA. Interactions among *Heliothis virescens* larvae, cotton condensed tannin and the CryIA(c) $\%_0$-endotoxin of *bacillus thuringiensis*. Journal of Chernical Ecology. 1993;**19**:2485-2499

[56] Howell CR, Bell AA, Stipanovic RD. Effect of aging on flavonoid content and resistance of cotton leaves to *verticillium* wilt. Physiological Plant Pathology. 1976; **8**(2):181-188

[57] Liu Q, Luo L, Zheng L. Lignins: Biosynthesis and biological functions in plants. International Journal of Molecular Sciences. 2018;**19**(2):335. DOI: 10.3390/ijms19020335

[58] Huang JQ, Chen XY. Removing the toxic enantiomer from crop products with selective gene editing. Naure Plants. 2023;**9**:513-514. DOI: 10.1038/s41477-023-01380-6

[59] Bell AA, Stipanovic RD. Biochemistry of disease and pest resistance in cotton. Mycopathologia. 1978;**65**:91-106

[60] Stipanovic RD, Bell AA, Mace ME, Howell CR. Antimicrobial terpenoids of *Gossypium*: 6-methoxygossypol and 6, 6′-dimethoxygossypol. Phytochemistry. 1975;**14**:1077-1081

[61] Zhao T, Xie Q, Li C, et al. Cotton roots are the major source of gossypol biosynthesis and accumulation. BMC Plant Biology. 2020;**20**:88. DOI: 10.1186/s12870-020-2294-9

[62] Punit M, Phundan S, Narayanan S, Mohan P, Singh P. Variability for gossypol glands in upland cotton (*Gossypium hirsutum* L.). Advances in Plant Sciences. 1991;**4**:165-170

[63] Altman DW, Stipanovic RD, Bell AA. Terpenoids in foliar pigment glands of a, D, and AD genome cottons: Introgression potential for pest resistance. Journal of Heredity. 1990;**81**: 447-454

[64] Adams R, Morris RC, Geissman TA, Butterbaugh DJ, Kirkpatrick EC. Structure of gossypol. XV. An interpretation of its reactions. Journal of American Chemical Society. 1938;**60**: 2193-2204

[65] Campbell KN, Morris RC, Adams R. The structure of gossypol. I. Journal of

American Chemical Society. 1937;**59**: 1723-1728

[66] Smith FH. Biosynthesis of gossypol by excised cotton roots. Nature. 1961; **192**:888-889

[67] Cook OF. Weevil-resisting adaptions of the cotton plant. US Department of Agriculture, Bureau of Plant Industry Bulletin. 1906;**88**:1-114

[68] Parrott WL. Plant resistance to insects in cotton. The Florida Entomologist. 1990;**73**:392-396. DOI: 10.2307/3495457

[69] Jenkins JN. Assessment of pest management for glandless cotton: Agronomic viewpoint. In: Glandless Cotton, its Significance, Status, and Prospects. Proceedings of a Conference on Glandless Cotton; 13–14 Dec. 1977; Dallas. New Orlean: USDA ARS; 1978. pp. 110-117

[70] Jan M, Liu Z, Guo C, Zhou Y, Sun X. An overview of cotton gland development and its transcriptional regulation. International Journal of Molecular Science. 2022;**23**(9):4892. DOI: 10.3390/ijms23094892

[71] Ye K, Teng T, Yang T, Zhao D, Zhao Y. Transcriptome analysis reveals the effect of grafting on gossypol biosynthesis and gland formation in cotton. BMC Plant Biology. 2023;**23**:37. DOI: 10.1186/s12870-022-04010-z

[72] Maga JA, Lorenz K. Gas-liquid chromatography separation of the free phenolic acid fractions in various oilseed protein sources. Journal of the Science of Food and Agriculture. 1974;**25**:797-802

[73] Veksler NA, Smirnova LS, Abduazimov KA. Isolation and study of individual fractions of dioxane lignin from cotton-plant stems. Chemistry of Natural Compounds. 1978;**14**:98-100

[74] Fuller MGA. Using Weed-Suppressive, Chromosome Substitution (CS) Cotton as a Supplemental Weed Management Strategy [Thesis]. Mississippi State: Mississippi State University; 2021

[75] Radcliffe JD, Czajka-Narins DM, Imrhan V. Fatty acid composition of serum, adipose tissue and liver in rats fed diets containing corn oil or cottonseed oil. Plant Foods for Human Nutrition. 2004;**69**:73-77

[76] Wood R. Comparison of the cyclopropane fatty acid content of cottonseed varieties, glanded and glandless seeds, and various seed structures. Biochemical Archives. 1986;**2**: 73-80

[77] Gorham J. Glycinebetaine is a major nitrogen-containing solute in the malvaceae. Phytochemistry. 1996;**43**: 367-369

[78] Wendel JF, Brubaker CL, Percival AE. Genetic diversity in *Gossypium hirsutum* and the origin of upland cotton. American Journal of Botany. 1992;**79** (11):1291-1310

[79] Bell AA, Stipanovic RD, Elzen GW, Williams HJ Jr. Structural and genetic variation of natural pesticides in pigment glands of cotton (*Gossypium*). In: Waller GR, editor. Allelochemicals: Role in Agriculture and Forestry. ACS Symposium Series 330. Washington DC: American Chemical Society; 1987. pp. 477-490

[80] Clancy MV, Mamin M, Flückiger G, Quijano-Medina T, Pérez-Niño B, Abdala-Roberts L, et al. Terpene chemotypes in *Gossypium hirsutum* (wild cotton) from the Yucatan peninsula,

Mexico. Phytochemistry. 2023;**205**: 113454

[81] Wang Q, Eneji AE, Kong X, Wang K, Dong H. Salt stress effects on secondary metabolites of cotton in relation to gene expression responsible for aphid development. PLoS One. 2015; **10**(6):e0129541. DOI: 10.1371/journal.pone.0129541

[82] Opitz S, Kunert G, Gershenzon J. Increased terpenoid accumulation in cotton (*Gossypium hirsutum*) foliage is a general wound response. Journal of Chemical Ecology. 2008;**34**(4):508-522. DOI: 10.1007/s10886-008-9453-z

[83] Mbiza NIT, Hu Z, Zhang H, Zhang Y, Luo X, Wang Y, et al. *GhCalS5* is involved in cotton response to aphid attack through mediating callose formation. Frontiers in Plant Science. 2022;**13**:892630. DOI: 10.3389/fpls.2022.892630

[84] Kolahi M, Faghani E, Kazemian M, Goldson-Barnaby A, Dodangi S. Changes in secondary metabolites and fiber quality of cotton (*Gossypium hirsutum*) seed under consecutive water stress and in silico analysis of cellulose synthase and xyloglucan endotransglucosylase. Physiology and Molecular Biology of Plants. 2021;**27**(8):1837-1857. DOI: 10.1007/s12298-021-01033-y. Epub 2021 Aug 6

[85] Zhang X, Wang G, Xue H, Zhang J, Wang Q, Zhang Z, et al. Metabolite profile of xylem sap in cotton seedlings is changed by K deficiency. Frontiers in Plant Science. 2020;**11**:592591. DOI: 10.3389/fpls.2020.592591

[86] Hafeez M, Liu S, Jan S, Ali B, Shahid M, Fernandez-Grandon GM, et al. Gossypol-induced fitness gain and increased resistance to deltamethrin in beet armyworm, *Spodoptera exigua* (Hubner). Pest Management Sciences. 2019;**75**:683-693

[87] Fernández-Aparicio M, Masi M, Cimmino A, Vilariño S, Evidente A. Allelopathic effect of quercetin, a flavonoid from *Fagopyrum esculentum* roots in the radicle growth of *Phelipanche ramosa*: Quercetin natural and semisynthetic analogues were used for a structure-activity relationship investigation. Plants (Basel). 2021;**10**(3): 543. DOI: 10.3390/plants10030543

[88] McMichael SC. Hopi cotton, a source of cottonseed free of gossypol pigments. Agronomy Journal. 1959;**51**:630

[89] Rathore KS, Pandeya D, Campbell LM, Wedegaertner TC, Puckhaber L, Stipanovic RD, et al. Ultra-low gossypol cottonseed: Selective gene silencing opens up a vast resource of plant-based protein to improve human nutrition. Critical Reviews in Plant Sciences. 2020; **39**:1-29. DOI: 10.1080/07352689.2020.1724433

[90] Naeem MA, Khan MH, Nazish T, Bashir U, Midrarullah. Allelopathic potential of crop water extracts of *Gossypium hirsutum* on germination and growth of wheat (*Triticum aestivm* L.). Pakistan journal of weed science. Research. 2023;**29**:9-18

[91] Mierziak J, Kostyn K, Kulma A. Flavonoids as important molecules of plant interactions with the environment. Molecules. 2014;**19**(10):16240-16265. DOI: 10.3390/molecules191016240

[92] Iqbal N, Khaliq A, Cheema ZA. Weed control through allelopathic crop water extracts and S-metolachlor in cotton. Information Processing in Agriculture. 2020;**7**:165-172

[93] Yue Z, Tseng T-M, Lashley M. Characterization and deer-repellent

property of chrysophanol and emodin from sicklepod weed. American Journal of Plant Sciences. 2018;**9**:266-280. DOI: 10.4236/ajps.2018.92022

[94] Parrott WL, Jenkins JN, Mulrooney JE, McCarty JC Jr, Shepherd RL. Relationship between gossypol gland density on cotton squares and resistance to tobacco budworm (Lepidoptera: Noctuidae) larvae. Journal of Economic Entomology. 1989;**82**:589-592

[95] Hofte H, Whiteley HR. Insecticidal crystal proteins of bacillus thuringiensis. Microbiological Reviews. 1989;**53**:242-255

[96] Gill SS, Cowles EA, Pietrantonio PV. The mode of action of *bacillus thuringiensis* endotoxins. Annual Review of Entomology. 1992;**37**:615-636

[97] Altman DW, Stipanovic RD, Benedict JH. Terpenoid aldehydes in upland cottons. II. Genotype-environment interactions. Crop Science. 1989;**29**:1451-1456

[98] Awadalla OA, El-Refai IM. Herbicide-induced resistance of cotton to *verticillium* wilt disease and activation of host cells to produce the phytoalexin gossypol. Canadian Journal of Botany. 1992;**70**:1440-1444

[99] Duke SO, Paul RN, Becerril JM, Schmidt JH. Clomazone causes accumulation of sesquiterpenoids in cotton (*Gossypium hirsutum* L.). Weed Science. 1991;**39**:339-346

[100] John ME, Keller G. Metabolic pathway engineering in cotton: Biosynthesis of polyhydroxy-butyrate in fiber cells. Proceedings of the National Academy of Sciences USA. 1996;**93**: 12768-12773

[101] Gao L, Chen W, Xu X, Zhang J, Singh TK, Liu S, et al. Engineering trienoic fatty acids into cottonseed oil improves low-temperature seed germination, plant photosynthesis and cotton fiber quality. Plant and Cell Physiology. 2020;**61**(7):1335-1347

[102] Lin J-L, Fang X, Li J-X, Chen Z-W, Wu W-K, Guo X-X, et al. Dirigent gene editing of gossypol enantiomers for toxicity-depleted cotton seeds. Nature Plants. 2023;**9**:605-615. DOI: 10.1038/s41477-023-01376-2

[103] Miller RF. Structure of Gossypol [Thesis]. Urbana: University of Illinois; 1937

[104] Bell AA, Christiansen MN. Gossypol synthesis in chilled cotton tissue. Phytopathology. 1968;**58**:883

[105] Yue Z, Tseng T-M, Reddy KR. Cotton defense response to adjacent plants and allelopathy (in preparation)

[106] Mississippi State University Extension Service. What is the Recommended Seeding Rate for Cotton? Available from: https://extension.msstate.edu/content/what-the-recommended-seeding-rate-for-cotton [Accessed: December 1, 2023]

[107] Mississippi Weed Science Committee. Weed Control Guidelines for Mississippi. Publication 1532 (rev-03-23). Mississippi State: Mississippi State University; 2023. p. 265

Chapter 2

Enhancing Water Use Efficiency by Using Potassium-Efficient Cotton Cultivars Based on Morphological and Biochemical Characteristic

Muhammad Naeem Akhtar, Muhammad Waseem Akhtar, Ashfaq Ahmad Rahi and Tanveer ul Haq

Abstract

Pakistan's agriculture sector faces a 50% water shortage, impacting crops like cotton and the textile industry, thus affecting the economy. K-efficient cotton cultivars, which withstand dry conditions through morphological and biochemical traits, can conserve moisture. K-efficient cotton cultivars can be used as donors of key K acquisition traits in breeding programs to develop promising cotton varieties with enhanced yields for the low K environments and can also be recommended for general cultivation to improve cotton production on a sustainable basis. This chapter aims to contribute to enhance water use efficiency (WUE) on the physiological basis to develop strategies for K-efficient cotton cultivars for improvement. Strategies involve regulating transcription factors to improve K-uptake efficiency, increasing root volume through lateral roots and root hairs, and enhancing K-uptake via channels and transporters, ultimately boosting WUE. Molecular breeding programs can leverage K+-associated QTLs to develop high K+ use efficiency cultivars. Physiological processes affecting WUE are discussed alongside factors influencing their contributions, acknowledging the complexity across crops, environments, and nutrients.

Keywords: K-efficient cultivars, gas exchange traits, morphological traits, water use efficiency, potassium use efficiency

1. Introduction

Cotton, which is a prominent commercial crop in over 30 nations, is produced mostly in warmer areas and has large shareholdings in China, India, the United States, and Pakistan [1]. In 2014–2015, the top five cotton-productive nations were China, India, Pakistan, and Brazil, with 6.5, 5.4, 3.5, 2.3, and 1.5 M, respectively [2]. As a glycophyte, cotton is more tolerant than other main crops to abiotic stressors. However, e extreme conditions, such as drought, affect cotton growth, productivity, and fiber quality [3]. Cotton production is predicted to decrease owing to drought stress, according to a news release issued by the United States Department of Agriculture [4].

Similarly, cotton output decreased by 34% in Pakistan to only 9.68 million bales compared with 14.4 million bales in 2012, due to drought and high weather [5]. Other crops were harmed by drought in addition to cotton, as about 67% of crop losses in the United States were caused by drought stress in the last 50 years [6].

Water use efficiency (WUE) can be increased by reducing soil water evaporation and converting more moisture into biomass production through crop residue management, mulching, and target-based irrigation. Appropriate arrangement and improved water use is a possible measure for improving yield under water stress, and it assesses the way and depth of water application, and whether it was used at optimum level by the crop. Improved irrigation water use efficiency can be achieved by approving the best management practices of irrigation [7].

Potassium (K) is affianced in all the physiological developments of the plant, which need water. K supports water carrying and mineral translocation for the entire plant through the xylem. Under K-deficiency, the movement of minerals such as (NO_3^-), (PO_4^{3-}), (Ca^{2+}), (Mg^{2+}), and amino acid uptake is abridged. One option is to use the normal rate of K application can support the plant to alleviate the effect of the water-deficit through better water use efficiency (WUE), which was related to the lower leaf ET [8]. Secondly, the option is to use K-efficient crop cultivars that can reduce the effect of water shortage and improve WUE without compromising the yield [9]. This can be achieved by the K-efficient cultivar by special proteins present in the cell membrane. These are called transporters and channels. Based on their affinity for K^+, K^+ transport components can be classified as high-affinity components (transporters), which are active at a low concentration of external K^+, and low-affinity components (channels), which are active at a higher concentration, usually at more than 0.3 mM external K^+ [10]. During water shortage conditions, root regulate their water and ion uptake by modifying these proteins and channels to cope with the water shortage [6, 11–32]. It is, therefore, K-efficient cotton cultivars that could perform better for sustainably cotton production under arid climatic conditions.

1.1 Cotton is drought-sensitive crop

Millions of people worldwide are involved with cotton directly or indirectly. Cotton is a severely susceptible crop under water stress conditions that reduces productivity [33]. Earlier, water stress reduced cotton growth and cotton yield due to lowered flowering and retention of the boll. Cotton is classified as a low drought-sensitive crop, according to FAO data; however, drought stress is dreadful when cotton is at the flowering and bolls formation stages [34]. In Pakistan, the cotton crop is mainly grown in arid and semiarid climatic conditions where annual rainfall is less than 250 mm. In the future, therefore, cotton productivity is estimated to drop due to severe water shortages, irregular patterns of rainfall, and other environmental impacts [35].

Drought has broad and variable effects on cotton, such that exact financial figures are hard to calculate. In 2008 and 2009, world production of cotton was extremely low and in 2009 stocks declined significantly. As a result, in 2010 and 2011, cotton prices increased, causing a 10% drop in cotton consumption in 2011. From 2010 to 2013, the output of cotton was more than demand, but from 2011 production declined significantly from 2014 to 2015 by 6.5 percent, while consumption grew by over 6.5 million bales per year. We thus need to set cotton production and consumption strategies. Furthermore, because too unpredictable situations in the future, stressful cotton types must also be produced. The focus must, on the other hand, not only be placed on the diversity of stress-tolerant of cotton, although plant survival is highly essential

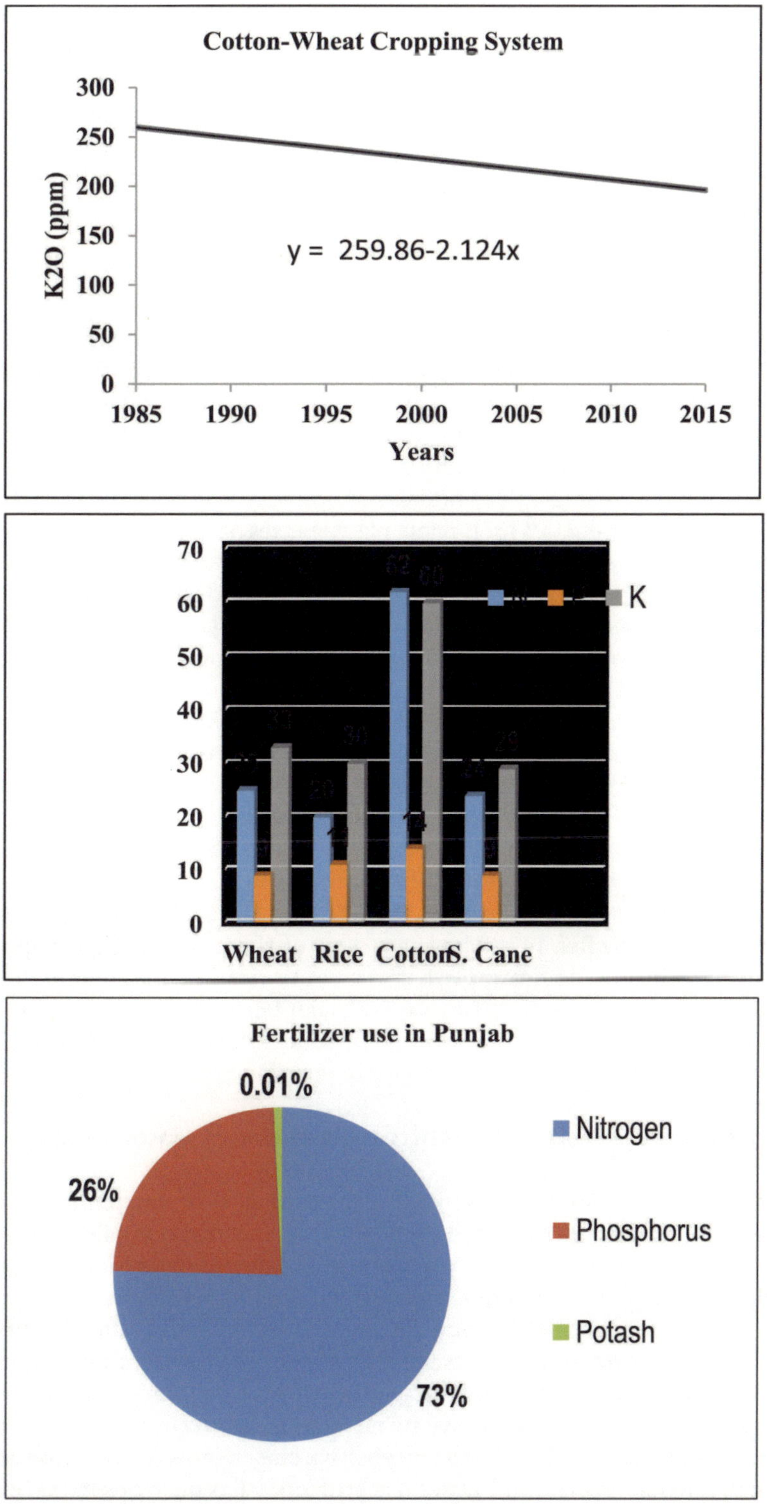

Figure 1.
K status and crop removal in Pakistani soils [11].

in the early stages of growth. It is widely recognized that improvements in yield and the stability in yields in cotton cultivations are vital for the expanding worldwide population under normal and drought conditions. Despite the complexities of the process of drought tolerance in cotton, the knowledge of the drought tolerance mechanism has been tremendously advanced. The drought-tolerant type of cotton might lead either to morpho-physiological, biochemical, and molecular modifications by nature or genetic engineering.

1.2 Current scenario of K-deficiency

The soil contains on average 2% K; however, soil K may be quite low in older or drained soils. Soil K is usually available in four pools: 0.1–0.2% of the soil solution, 1–2% of the exchangeable K, 1–10% of the nonexchangeable K, and 90–98% of the soil mineral solution [36]. Plant roots collect K from variable pools and dynamically adjusted K solution for soil [36, 37]. The soil K in a solution pool of soil is filled by potassium released from interchangeable areas [36]. The nonexchangeable K may be found in the range of 2: 1 and 2: 1: 1 type clay minerals [37]. Some plant species have been discovered to utilize K from an unchangeable pool, such as sugar beet (Beta vulgaris L.) with exudates, which results in a K-influx of 7–20% higher than that for wheat and barley roots [36].

The deficiency of K in Pakistani soils is about 35% [38], which is continuously increasing due to the low use of K (<1.0 kg K_2O ha^{-1}) as compared to other countries of the world (15 kg K_2O ha^{-1}). In the year 2007–2008, about 43% of Pakistani soils were deficient in K, as reported by Ref. [39], but currently, it reached up to 90% and needs K fertilization for better crop production. Cotton requires K_2O @150 kg ha^{-1} that almost equal to nitrogen from the soil, to get average crop yield [40] and is required @ 4.5 kg ha^{-1} day^{-1} during peak bloom by cotton [41]. Despite high-K requirements, K-use in cotton production is very limited in Pakistan. The total content of K varies from 50 to 150 mg kg^{-1} in the soil across the Punjab province. Adequate K nutrition upgrades the proficiency of photosynthetic rate and development of plant roots, as investigated by Ref. [42]. However, most of the farmers of developing countries, including Pakistan, do not apply K-fertilizer to crops due to their high-prices. Under this situation, the identification of K-efficient crop cultivars can improve production and reduce the demand for K fertilizers in the country (**Figure 1**).

2. Biochemical modifications in cotton plant under water stress

In many plant activities, water is vital for the transport of nutrients, chemical and enzymatic reactions, cell growth, cell division, and transpiration [43]. Drought stress reduces plant development due to disruptions in the plant's main biochemical and physiological systems [44]. Drought stress affects root penetration, and stem elongation and increases water use efficiency. The leaf water potential, rate of transpiration, and leaf temperature are important traits that affect the growth of plants under water-deficit conditions. Although all stages of cotton development are affected by water stress, however, the reproductive phase, that is, flowering and boll development, are generally accepted as the most sensitive stages [45]. A very close relationship between nutrient uptake and water use efficiency has been observed in many crops. Therefore, nutrient uptake was reduced during drought stress, particularly N and K in cotton. Similarly, K contents in plant tissue also decrease due to drought stress and an overall reduction in nutrient uptake and there utilization [43].

Root growth is important for plant development as water and nutrients are mainly taken up via roots, whereas root elongation is affected by drought stress which ultimately limits crop yield [20]. Under severe drought stress, the elongation of plants may be inhibited by interrupted water flow from the xylem to the surrounding cells, which impairs mitosis and cell expansion, resulting in a reduction of root growth, nutrient uptake, plant height, and leaf area as well as plant growth [46]. In cotton, under water stress conditions chlorophyll contents decreased which reduced photosynthesis rate, sugar production, and pigments including chlorophyll *a*, chlorophyll *b,* and carotenoids [47].

There is a root-to-leaf transduction of chemical signals caused by water-deficit stress, through the production of abscisic acid (ABA), which results in the closure of stomata. Under drought stress, ABA promotes stomata closure to reduce the transpiration rate [48]. When water potential is low in plants, ABA also stimulates root growth and inhibits shoot growth. Once the soil water availability is reduced, the amount of ABA in the xylem increases, and consequently, ABA concentration in different parts of the leaf is increased. As ABA is directly associated with stomatal conductance, it reduces net CO_2 intake and decreased photosynthesis [48]. Existing cotton cultivars vary in tolerance to abiotic stresses, such as drought; therefore, maintenance of optimum plant cell turgor is an indication of the drought tolerance potential of a cultivar [49]. Similarly, a change in carbohydrate metabolism with an increase in glucose concentration in leaves and sucrose concentration in pistils of white flowers of cotton was observed under drought stress [50].

Ascorbate peroxidase activity is reported to increase whenever plants undergo in drought stress as reported in cotton grown under drought stress but glutathione metabolism levels were not changed [51]. However, the investigation of antioxidant activity in cotton plants exposed to drought stress is still controversial and is not well understood and superoxide radicals (SOD) or catalase (CAT) activities remained unchanged under drought stress [52]. It has been shown that CAT activity is not affected by drought stress, while, the Ascorbate peroxidase (APX) and SOD activity was increased [53]. In most plants, osmoregulation reduces the water potential of cells, thus increasing the gradient for water flow in the cell to maintain cell growth [54]. The maintenance of cell turgor contributes to ongoing physiological processes such as stomatal conductance and photosynthesis, as reported by Ref. [21, 55]; however, information on osmotic regulation in modern cotton varieties under drought stress is still lacking.

Yield is essentially the precise integration of the many systems of physiology. Drought stress negatively affects most of these physiological systems. The negative effects of shortage of water on yield depend mainly on the intensity of stress and the growth stage of plant life. In key crops, significant drought stress losses have been observed. The low moisture caused by parenthesis lowered the time to anthesis and reduced the time to fill the grain with cereal by anthesis [56]. Exposure of plants to drought stress can lead to total sterility of pearl millet (*Pennisetum glaucum* L.), typically owing to disrupted mobility in the assimilated ear [57]. The drought may be caused by many causes such as decreasing photosynthetic rate [58], disturbing assimilate partitioning [44], or inadequate growth of a prominent leaf [59]. Maize was significantly reduced in returns when exposed to drought conditions at the tasseling stage [14]. In the same way, the production and abortion of the developed bolls in cotton under drought circumstances were significantly reduced, which eventually affected the lint yield [60]. In drought environments, there was also a substantial drop in barley grain production (*Hordeum vulgare* L.), largely due to lower viable grain tillers and grain with less than 1000 seeds [61]. Drought stress in the blooming stage produced more than a 50% decrease in the seed output by the exposition of pigeon peas (*Cajanus cajan* L.) [62].

Drought has a primary influence on the plants, namely poor germination and impaired planting. Several studies have demonstrated detrimental effects on germination and seedling development of drought stress [44, 63]. In major crops of fields, including pea (*Pisum sativum* L.), alfalfa (*Medicago sativa* L.), and rice (*Oryza sativa* L.) under drought stress, drought stress was observed to reduce germination, early sowing growth, root and shoot dry weight, hypocotyl ate length, and vegetative development [64–66]. Plant growth is achieved largely through the division of cells, expansion, and differentiation. Drought affects mitosis and cell elongation leading to poor growth [39]. Drought restricts the cell development process, largely owing to turgor loss [67]. Water limitation leads to cell elongation, largely because of the inadequate passage of water from the Xylem to the next cells [68]. Drought also reduces the number of leaves and the size of each leaf. The leaf growth usually depends on the turgor and the availability of assimilates. Reduced turgor and slower photosynthesis rate in circumstances of drought stress particularly restrict the growth of the leaf [59]. Fresh and dry weight in the water limitation conditions are likewise reduced significantly [69]. Plant height, leaf size, and, stem diameter were diminished significantly in maize under water-limiting conditions [70]. Ref. [71] found in another investigation that the bioaccumulation of maize has decreased considerably under dry conditions at different stages of growth.

Certain variables, such as the leaf water, leaf and canopy temperature, transpiration rates, and stomatal conductivity, impact water relations. Drought stress disrupts all of these processes in plants, but stomatal conductance is particularly impacted [44]. Drought conditions, which eventually raised the leaf and canopy temperature, showed a substantial decrease in the leaf water potential and transpiration rates [72]. The efficiency of the dry matter ratio accumulated with the water used is another essential characteristic of plant physiology control. Efficient wheat crops are more efficient at using water during drought conditions [73]. This improvement in the efficiency of water usage is large because the dry matter accumulates by absorbing less water due to the closure of the stomata and less transpiration. When subjected to an early season water scarcity, a decreased water efficiency in potatoes (*Solanum tuberosum* L.) was found, and eventually, biomass buildup and output were low [74].

3. Structural and physiological responses of the cotton plant under drought condition

Drought leads to a variety of changes in cotton plant growth and function. Drought, for example, seriously impedes many physiological processes regulating fiber quality and foliage output [45]. Mechanisms for drought resistance in plants consist of four groups, namely recovery, prevention, tolerance, and drought exhaust [75]. Water stress avoidance supports essential functions, such as stomatal control, in the event of moderate drought. The capacity of flora to undergo serious dehydration with osmotic adjustments and osmoprotectants is drought tolerance [76]. Plants have been developed to control the time of development to prevent moisture stress [77]. The capacity of plants to resume their development following drought damage is the recovery of drought. In the cotton area, biochemical, physiological, and molecular drought stress methods are examined in the preceding sections, as shown in **Figure 2**.

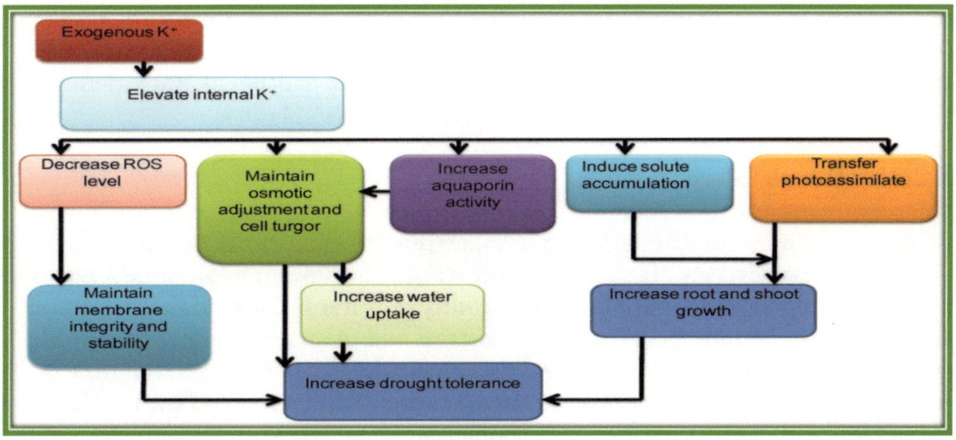

Figure 2.
Drought tolerance mechanism of potassium [12].

Plant roots are important for sensing and responding in direct contact with soil water and nutrients to different external environmental stimulation systems. Because the root structure configuration of dry soils is difficult to gather, little data on change in root systems are available in drought, and most research on cereal crops is done. Plant roots respond to the surface moisture changes, that is, water shortage at the top of the ground leads to deeper root penetration, whereas the upper layer of surplus water lowers the root penetration [78] by up to 3 m.

The growth of root rates for predicting agricultural production losses in cotton crops is widely used. Inadequate soil humidity inhibits root growth and development and so affects the working of the aerial portions [78]. Water shortage in the topsoil results in deeper root penetration for a more extensive moisture and nutrient exploration, whereas excess water in the top floor produces a decreased root penetration [78]. Drought decreases the formation of over ground biomass via a decrease in the density of root, root mass, and root lengths [79]. However, characteristics such as hydraulic conductance and plant allometry are of significant interest to scientists in the process of drought resistance. The rooting system with many short and thin side roots enables the acquisition of oxygen and soil nutrients to a wide root surface area, compared to the dispersed root system [80]. The Fine Root System drives soil activities such as carbon cycling, sequestration, nutrient flows, structural stability, and soil microbial activity [81]. For adjusting to drought, increased length of root and soil proliferation are desired characteristics. The growth and penetration of roots depend on external partial oxygen pressure inside the root region [82]. Mild drought stress may increase root elongation during the early stage but root morphological and physiological activity is severely impeded by long-term water stress disruptions [79]. Finally, deeper root insertion makes it possible for the plant to explore deeper water and nutrients in the soil. Therefore, it is necessary to promote the dispersion of vertical roots to improve crop growth and dry stress development. All root characteristics may be relevant for drought stress, but researchers have been particularly interested in hydraulic conductance and plant algometry. Different researchers have examined the probable origins of drought stress [6]. More abundant (greater radicular density) and deeper soil root systems often serve as desired features to adapt to drought. For one case, [79] reported that mild and initial drought stress improves cotton root length, but a long-term water shortfall reduces root activity in comparison to control plants. Transgenic cotton plants, with a stronger root structure than in the natural, were more resistant to drought stress in another trial [83]. Arabidopsis also harbored transgenic cotton plants that had a highly developed root gene, in addition to other characteristics tolerant to drought 1/homodomain glabrous 11 (AtEDT1/HDG11) [84].

3.1 Cotton lint yield

Lint yield is a complex integration of many physiological processes in cotton crops; most of them are severely affected by water stress. The development of new nodes in a cotton plant is dependent upon the availability of water because of the indefinite habit of growth. Duration and intensity of stress and plant development are linked to the detrimental effects of moisture stress on yield. Drought terminal significantly restricts the production of cotton by limiting carbon absorption and the buildup of biomass [85]. Inhibited synthesis of carbohydrates together with the depletion of the storage reserves (i.e., starch) owing to continuous breathing [86]. Ref. [87] reported that as a result of drought, reproductive structures and boll size decreases are induced [88]. Increased fruit and leaf abscissions might be linked to a final output loss for drought-stressed cotton crops [60]. In short, the loss of cotton production is closely linked to physiological and morphological plant stress processes.

3.2 Fiber quality

The cotton farmers are mostly focused on fiber quality, both as the fiber characteristics directly influence the fiber output and enhance the spinning processes [89]. The quality of the fibers is combined with the length of the fiber, fiber fineness of the wall, the strength of the fiber, the elasticity of the body fiber, naps (fiber nodules), short fiber index, uniformity index (fibers fitted in the spinning process), color grade and reflectiveness (fibers brightness) [19]. Fiber quality features are numerical and regulated by a variety of phenotypically significant and minor genes [90]. Lint quality is directly affected by water during fiber cell formation [91]. As a complex phenomenon with various morphological and physiological characteristics is connected with drought tolerance in the plant [92], the breeding of enhanced fiber quality characteristics under moisture stress is difficult [93]. The discovery for irrigated and water deficiency environments of stable quantitative traits loci (QTL) might therefore allow the molecular breeding of cotton genotypes with enhanced fiber quality and yield parameters. The use of abundant DNA markers for the cotton genome is necessary for the QTL, genetic diversity, and structural analysis [94].

Several QTLs have been highlighted in upland cotton on fiber yield features [95, 96]; however, there has been less emphasis given to identifying QTLs in terms of fiber quality under drought [97]. The inter-specific cotton plants F2 and F3 [98] are generated from a crossing between inbred lines *G. Siv'on* and *G. hirsutum* cv. CV F-177 barbecue. QTL (13 and 33) reported for the 16 QTLs, covering plant production, physiology, and fiber quality, under situations of well-watered and water deficiency. Ref. [28] discovered 79 QTLs combined with fiber quality features in the F2 and F3 generations which are derived from Siv'on and *G. hirsutum* cv. Barbadense cv F-177 underwater conditions are irrigated and deficient. Seventeen of the 79 QTLs discovered were moisture-specific conditions whereas just two were well-watered. In the F2 population, the mapped physiological, yield, and plant structural characteristics in the cross between G were created [97]. Hirsutum cv. Hirsutum cv FH-901 (sensitive to drought) and *G. hirsutum* cv. Hirsutum cv RH-510 RH-510 (drought-tolerant).

A total of seven QTLs, three and two of which are water-limited and well-watered, were identified accordingly. These germplasm panel QTL analyses included a large range of relevant alleles that may be detected on *G. hirsutum* lines with various genetic information. In this work, 177 single SSR markers were utilized in a panel of 99 upland genotypes of cotton for the detection of significant quantitative trait loci (QTL) associated with 11 quality fibers and structural plant characteristics. The quality of fiber and the structural features of plants under water and water deficits were examined in another investigation. GLM analysis revealed that 74 and 70 QTLs were found under well-watered and limiting water situations, respectively. The MLM found 7 and 23 well water and water shortage QTLs, respectively [19].

For example, efforts have been undertaken to discover the particular fiber gene and its activities for improving fiber quality for characteristics of significant interest in cotton fiber. Cotton genomics promises to increase the understanding and systematic utilization of fundamental plant biology to improve cotton fiber quality, and cotton functional genomics. However, it is laborious to determine the activities of cotton genes that a quick pace has not been evaluated [15]. Actin cytoskeleton [99], polysaccharide biosynthesis, [2]), and the related genes are expressed in distinct routes for fiber formation. Of these, few are mostly present in the production of fiber [100], secondary biosynthesis cell wall [25], and fiber-extension [101]. A protodermal cotton gene 1 (GbPDF1) has now been expressed through the HDZIP2ATATHB2 core element at the fiber initiation stage [102]. In developing fibers, alpha expansion genes (GhExp1) encode a cell wall protein and govern the loosening of the cell wall [103]. Ref. [104] demonstrated the interruption of the fiber elongation and SuSy's role in osmosis control by the antisense removal of the sucrose synthase (SusSy) gene. On the other hand, GhPRP5 was a negative regulator in fiber formation, with proline-rich protein genes coding [105]. During secondary cell wall biosynthesis, cellulose synthesis is an important step in the formation of cells. Many types of research have been carried out to examine how cotton fiber controls and maintains the strong irreversible carbon sink that has a secondary synthesis of wall cellulose [25]. The subsequent discovery of a novel isoform Sus (SusC) during the secondary wall development of cellulose in fiber was followed [25]. Most of the expressed genes are associated with fiber maturity with cellular respiration [106]. Many transcription factors encode genes, that is, the fiber development phase included featured families from MYB, C2H2, bHLH, WRKY, and HD-ZIP. Past studies reveal that fiber formation in upland cotton has demonstrated significant expression in MYB-related genes [107]. Expression analyses from six MYB genes indicated that GhMYB6 was high in fiber, with R2R3 MYB gene encoding factor "GhMYB109" expressed in the fiber elongation and initiation [61]. The RAD-like GbRL1 is significantly expressed in cotton ovules at fiber commencement [108].

The identification of fiber-related loci markers can have useful impacts on genetic adaptation required under scarce water situations to create sufficient fiber. Many types of research into gene expression were conducted to understand the formation of cotton fibers, which offers problems. First, in comparison analysis, the bulk of the differentially expressed genes are connected to differences across species rather than to fiber-related features. Secondly, G protein-coding gene sequences are being used. G and Ramondii. Arboreum may not be precise enough for tetraploid cotton genetic analysis. Thirdly, it is unknown if any of the expressed genes found in previous research have changed the sequence between a mutant in cotton fiber and its natural type. In this context, the viable candidates for innovative cotton research are only differentially expressed genes with sequence variants and co-location with desired fiber properties.

3.3 Photosynthesis

Drought stress causes stomatal closure, reducing CO_2 absorption and consequently affecting photosynthesis rate [29]. In certain situations, however, stomata behavior is not necessarily linked to the photosynthetic rate, although this has to be clarified [109, 110]. As the water-deficit steadily rises in cotton, photosynthesis is strongly influenced along with growth. For example, photosynthesis and transpiration were shown to be altered in cotton in dry conditions [53, 106]. Incidentally, immature cotton leaves were shown to be photo-synthetically more resistant than mature leaves to drought and heat. There was no reduction in net photosynthesis when immature leaves were exposed to high temperatures (37°C). By comparison, the photosynthesis of mature leaf net decreased 66% under equal conditions [27]. A reduced lint production was reported when net photosynthesis deteriorated under water shortage conditions in the first growing season in another field trial of cotton in two consecutive growing seasons. However, there were no changes in the production of drought-treated fields in the next growing season due to heavy rainfall [28]. These researches have shown that drought stress decreases cotton photosynthesis, which in turn impacts growth and return.

Photosynthesis is the principal source of agricultural productivity, which is adversely affected by situations of water scarcity. When the stomata closures respond to wetness, the leaf photosynthetic capacity is reduced and the dehydration of chloroplast and CO_2 into the leaf is reduced (**Figure 3**). For example, moderate moisture stress induces stomata closure by controlling transpiration to prevent water loss. This decreases stomata behavior and limits the concentration of intercellular CO_2 [78]. In serious drought, decreased stomata conductance and (nonstomatal) metabolic damage, such as restricted carboxylation, become important photosynthetic constraints [87]. Likewise, stomata behavior is not continuously linked to photosynthesis, but investigation is needed [110]. Drought can impair both photosynthesis and seriously

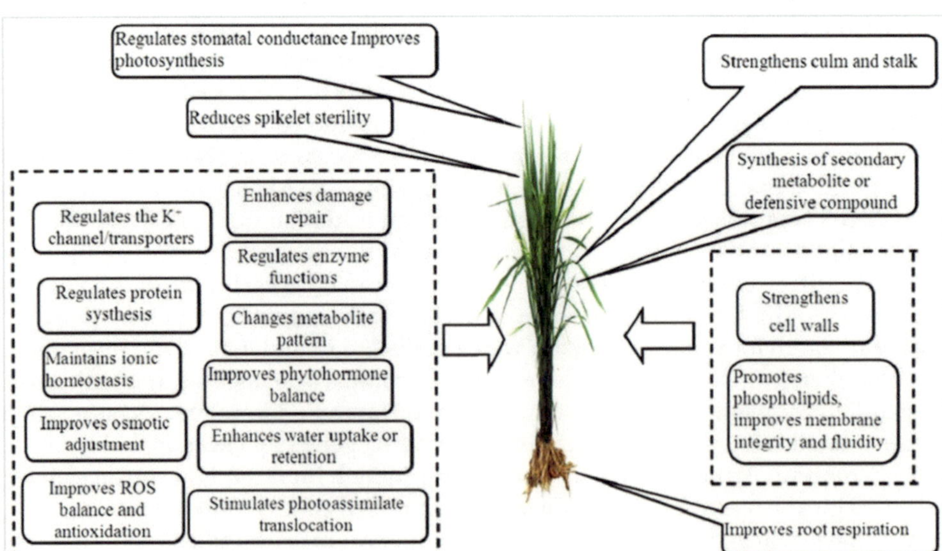

Figure 3.
Physiological role of potassium in plant.

influence transpiration and depend on the degree of drought and plant growth [111]. Under mature cotton leaves up to 66 percent decrease has been reported in water shortage situations about smaller leaves [27]. Overall, using K-efficient cultivars under reduced irrigation condition along with 50 Kg ha^{-1} K_2O improve the net photosynthetic rate by about 30% as compared to potassium nonefficient cotton cultivars [11].

3.4 Stomatal regulation

The major purpose of stomachs in plants is to control water loss through transpiration. In the event of a water-deficit-resistant plant, internal moisture preservation and fast stomach closure are critical. Water loss from cotton leaves is a major phenomenon in situations where water is in shortfall; however, the plants produce adaptations to survive under drought stress, that is, wilting and rolling leaves lead to less radiation interception and eventually lower water loss [75]. In general, a variety of xeromorphic characteristics increase drought tolerance, including a large cuticular epidermis, thicker and smaller leaves, smaller and denser stomata tissues, more epidermal trichomes in palisade, and a well-structured vascular bundle sheath [33]. In leaf gas exchanges between the leaf's internal cavity and its exterior surroundings, stomatal control plays an important function. The leaves of the plant disperse heat energy through three methods. Sensitive heat loss (lead and convection) and transpiration are processes such as re-radiation. Transpiration is the essential mechanism for flowering plants, for energy collection, and maintenance of cellular processes. With 90 percent of plant water loss occurring by transpiration [112], stomatal regulation plays a major function in ensuring a supply of water and nutrients for the vital physiological process. The first approach to reduce water loss in cotton cultivation in drought conditions is to close the stomata under excessive transpiration [103], in the eight maize hybrids showed that the portion of transportable soil water that the stomata began to close throughout a drying cycle was statistically different. Therefore, a potential biomarker for induced drought tolerance is stomatal conductance; however, a negative association is found between drought resistance and cotton stomatal conductance.

A significant feature in cotton plants under drought stress is the reduction of water loss through leaves. Wilting and rolling of the ground lead to less radiation and hence less loss of water [75]. Plastic materials frequently exhibit various xeromorphic characteristics, and have structures that encourage drought tolerance, such as thicker and smaller leaves, thicker cuticle epidermis, thicker palisade tissues, more dense stomata, a high proportion of palisades to the spongy perennial thickness, and a vascular sheath developed [33, 113]. In a gas exchange between tissues and the atmosphere, Stomata control plays a crucial function. It is one of the primary processes for the generation of energy and cell activity in plants. 90% of plant water losses are due to stomata openings [114]. The closing of the stomata is the first step in cotton to decrease water loss when the transpiration rate is very high. Stomata activity could be a possible indication of drought tolerance for cotton as the relationship between drought tolerance and stomatal performance is negative.

3.5 Osmotic adjustment

Osmotic adjustment is an adaptation technique for the acclimation of increased cell turgor and water retention after stress. Osmotic leaf adjustment is significantly associated with drought tolerance in other crop species. The osmotic adjustment of compatible solutes in the cytosol is carried out in plant cells in response to water stress. This lowers the cell's osmotic potential for cell turgor and cell growth. Compatible solutes such as proline, sorbitol, and glycine betaine are more soluble and do not interfere with metabolism in cells, even at high levels. Proline is a widely appropriate drought stress response in plants [24]. Proline accumulation in drought plants is, however, different and relies on cultivar and growing phase (e.g., proline accumulation in cotton ovaries was higher than in the leaves). Ref. [115] have suggested that during reproductive phases, the osmotic adjustment may be greater than in vegetative stages and may depend on tissue.

The water shortage affects the turgidity and osmotic balance of the cells at the cellular level. Osmotic adaptation is key to reducing the impacts of crop damage caused by drought. Mechanisms of plant protection also include osmoprotectants or osmolytes that control homeostasis after drought and cellular salinity stress. The effect of drought stress on osmotic balance is adverse and hence plants collect various organic and inorganic components to lower the osmotic potential of the dry weight [75]. Osmotic adjustment is involved in numerous organic compounds including amino acids (proline and glycine), sugar (trehalose and fructan), sugar alcohols (mannitol, sorbitol, D-monitor-monitor and polyamine (polyamine and betaine), polyols, ectoin, alkaloids, and inorganic ions known as osmoprotectants/osmolytes [75, 116]. Such solutes help protect proteins and membranes from harm owing to high concentrations of inorganic ions and oxidative damage caused by drought stress [31]. The exogenous use of osmoprotective agents (proline and glycine betaine) has proven to be beneficial in decreasing the deleterious effects of cotton drought stress [117]. Transgenic cotton plants were more drought-tolerant than controlling plants and were more photosynthesized, had greater relative water content, improved osmotic adjustment, reduced lipid membrane peroxidation, and less ion leakage [76]. AnnBj1 ectopic annexin gene expression improved the content of the proline and sucrose, which increased the tolerance of drought in cotton [118]. In addition, the overexpression of the GhAnn1 cotton annexin gene improved dryness and salt tolerance by boosting the activity of superoxide dismutase (SOD) [119].

Compatible solutes protect the proteins and membranes from damage due to high levels of inorganic ions and water-deficit oxidant damage [31], and salinity [6]. Foliar application of glycine betaine and proline could be a useful approach for increasing tolerance in cotton cultivations [117]. More drought tolerance showed in cotton plants more glycine betaine in accumulation. The promotion of physiological processes, such as leaf photosynthesis, relative water, improved osmotic adjustment, and low lipid stability, might thus increase crop performance under drought by transgenic/nontransgenic method [42]. For example, a rise in proline and sucrose content of the AnnBj mustard annexin gene in cotton led to a higher tolerance of drought [118]. In addition, GhAnn1 overexpression, annexin cotton gene, drought, and salt tolerance have been enhanced by the enhancement of superoxide dismutase [119]. Further osmotic adjustment research in reproductive organs is needed to completely understand this process in drought cotton plants.

4. Biochemical and molecular mechanisms of drought tolerance

Plants via morphological adaptation avoid a range of environmental stressors. The drought tolerance mechanism is connected with several biochemical, morph physiological and molecular processes. The hormonal interaction inside the plant body controls these activities intensely.

4.1 Abscisic acid (ABA)

ABA is a natural plant stress hormone; stress response, growth, and reproductive behavior in cultivated plants. Osmotic stress in plants is linked to drought and the low available water induced by the synthesis of ABA and adaptation mechanisms [22]. The abscisic acid production is triggered after the reception of stress signals by the plasma membrane, excluding the xanthoxin transition into ABA. This usually happens in the cytoplasm [120]. ABA is typically rooted and transmitted by vascular tissues to higher regions of the plant [111]. In cotton, ABA is detected and transmitted using ABA-dependent or ABA-independent transmission, whereas the former are essential actors for stress-responsive gene expression under many stressors, including osmotic pressures. Many receptors in the plasma membrane, cytosol, the envelope of chloroplast, and nucleus have been identified. The plants display a low ABA concentration in the nonstress environment; sugar nonfermenting protein kinase 2 (SnRK2) is suppressed by protein phosphorylation of 2C (PP2C). ABA enhances drought tolerance in cotton plants through the modulation of stress-related genes. ABA-induced overexpression of cotton genes GhCBF3 in Arabidopsis led to drought tolerance in transgenic lines, with greater levels of relative water, chlorophyll, and proline than wild type [121]. The AREB1 and AREB2 are more expression-level compared to the wild transgenic line, while the stomata aperture is lower when ABA is treated. Suggesting that, through the ABA signaling route, GhCBF3 may increase drought resistance.

4.2 Jasmonic acid (JA)

Plant phytohormone and its active by-products known as jasmonate are considered Jasmonic acid (JA). In fighting many biotic and abiotic stressors, it plays a crucial function. In addition, JA is related to improving root structures, tendril coiling, pollen generation, and fruit maturation [122]. Exogenous application of jasmonate has been found to enhance plant performance in drought environments [18, 123]. The mechanism and production of jasmonic acid signaling were intensively researched. The repressor protein jasmonate-zim (JAZ) plays an important function in the JA signalization pathway as the JA signaling switch. Jasmonate/Jasmonate-Zim (JAI3/JAZ), proteins that are not under stress or lack of JA, are associated with and eliminate various elements of transcription, including myelocytomatosis (MYC2). Nonetheless, the degradation of JAZ proteins, as illustrated above, occurs under deficiency water, producing the effect of active transcription factors, that is, MYC2, which regulates stress tolerance-related genes [32]. Plant hormones usually do not work in one path, but depend partly on each other in different periods to control the environment and development paths. In plants, signal transduction occurs, and many changes may be organized to adapt in a challenging way to harsh conditions [123].

4.3 Reactive oxygen species (ROS)

The fractional reduction of ambient O_2 is responsible for the formation of reactive oxygen species (ROS). The cell ROS consists of four categories: radical hydroxyl (HO·), radical anion superoxide (O_2-), peroxide hydrogen (H_2O_2), and singlet oxygen ($1O_2$). HO· and 1O2 are very reactive and may oxidize DNA and eventually causes cell death with RNA, lipids, and proteins [75]. The generation of ROS induces subcellular locations, namely the cell wall, chloroplast, nucleus, mitochondria, and plasma membrane [124]. Drought production increases such as a reduction in CO_2 fixation leading to a decrease in NADP+ reconstruction during the Calvin cycle. This lowers the activity of the photosynthesis chain. Moreover, the Mehler reaction in electrodes may potentially increase the generation of ROS during photosynthesis by too many electrons leaking to O_2 [125]. By donating an electron in photosystem-I, the Mehler reaction decreases O_2 to O_2. O_2- via superoxide dismutase, which can be converted into water by ascorbate peroxidase, may be changed into hydrogen peroxide. It is difficult, however, to evaluate ROS levels of those created by photorespiration during the Mehler reaction. The photo-respiratory route also increases moisture stress, especially if RUBP oxygenation is strong because of the partial fixing of CO_2. Some 70 percent of overall humidity stress generation H_2O_2 takes occurs by photorespiration [125].

Plants have sophisticated systems to check the ROS redox homeostasis, to avoid extra ROS in cells. Changes in the metabolism of antioxidant enzymes may impact drought resistance in cotton plants. Plants have created antioxidant mechanisms to keep growing. This system consists of enzyme and nonenzyme additives. These enzymes include dismutase of superoxides, ascorbate peroxidase, and peroxides of guaiacol, reductase monodehydroascorbate, catala, reductase, and glutathione reductase of dehydroascorbate. Nonenzymatic components include reduced ascorbic acid (AA), flavonoids, carotenoids, proline, glutathione, and (GSH). Both components operate together to break ROS [126, 127]. The Halliwell Asada route detoxifies the H_2O_2 with the ascorbate peroxidase along with NADH MDAR, and GR [127]. In MDHAR, Ascorbate decreases MDHA. However, 2 MDHA molecules can be converted into MDHA and dehydroascorbate without being enzymatically reduced to ascorbate via the NADH and GR cycles [128]. The presence of NADPH reduces glutathione (GSH) via GR oxidation. Glutathione reductase activity rises with humidity stress, retaining oxidized and decreased levels of glutathione [129]. If oxidative signals and/or losses occur, the balancing of antioxidant activity and ROS generation determines whether [26]. The antioxidant capacity of various cotton cultivars affects the potential resilience to dry conditions. The moisture stress in cotton causes ROS formation; however, the ROS scraping process may also be improved and maintained by APX and GR activities [130].

The use of nutrients (Zn) was shown to reduce oxidative damage to cotton caused by polyethylene glycol (PEG). This raises the amount of CAT, APX, SOD, and enzyme-free antioxidants [52]. Increasing GR activities and better lump-sum levels have been reported in [130] drought-tolerant (CCRI-60). The CCRI-60 was capable of scavenging free radicals and protecting the plants against severe circumstances, in comparison to the sensitive (CCRI-27). This indicates better development and increased drought stress resistance. GbMYB5 down-control in Gossypium barbadense has resulted in decreased activity with antioxidants such as CTA, peroxidase (POD), SOD, and Glutathione S-transferase (GST) [30]. Further research is, however, necessary to discover genes in drought-resistant cotton cultivar pathways associated with the antioxidant enzyme. The use of Zn and K supplies can also help strengthen the cotton plant's antioxidant system [131].

5. Potassium as a drought stress ameliorating tool

Increases the leaf photosynthetic of crops, crop maturity, and high-quality fibers among all plant nutrients, K has a key role in increasing vigorous cotton development [23]. Due to its negative effects on cotton's surface of leaves of plants, photosynthesis, and production of biomass, K was shown to reduce yield and fiber quality [132], because K plays a major role in cotton growth, development, and quality fibers.

In addition, excess or insufficient K is vital to the normal function of the plants and also for the development of plants [9] in the growth medium to maintain the K level and its relation with other critical plant nourishment (particularly sodium (Na)). The management of K-fertilizer is, therefore, beneficial to enhance plant growth. Adequate K supplies influenced the overall growth of cotton substantially [133] because of their crucial function in the generation of biomass [134], an increase in its surface area, and the synthesis of photosynthesis [135].

Cotton is more vulnerable to low K and the most sensitive to K fertilization than other plants [81]. Ref. [133] He also found varied responses from conventional and Bt cotton cultivars to low K levels. In biomass output, four cotton cultivars are significantly different from K in a field test. Likewise, nutrient-efficient genotypes culture was proposed as a key strategy to enhance fertilizer efficiency [37]. By decreasing the usage of artificial fertilizers in agriculture, potassium intake and effective use of soil nutrients are anticipated to have good environmental benefits. The genotypes which are successful in K-uptake were proposed to have a broader root surface. This can lead to additional root-to-soil moves to maintain a broader root gradient. This might further enhance K translocation to different plant organs, which could maintain the optimum cytosolic K+ content. The enhanced selectivity from K to Na is the fundamental mechanism of better efficiency of use [37].

A sufficient supply of cotton K is not only necessary for water relations but also for improving the water use efficiency of (WUE) which can enable plants to live under drought stress [135]. Observed that WUE in cotton was enhanced by K but with cotton cultivars, this impact was different [136] shows that not all plants or K levels may have a favorable influence on every element of water in cotton stomach conductivity by fertilization with K under water-starved conditions.

Drought stress yields are a good tolerance to K applications, and they increased fiber yield and quality when compared with a well-watered control, under drought stress treatments [137]. Furthermore, it was observed that using K foliar spray, the micronaire, an indirect measure of fiber coarseness, was enhanced by 0.32. Drought stress saving the effects of drought stress and enhanced output was the use of foliar K. Although the reaction of cultivars was varied, the use of foliar K in the drought treatments increased fiber output and quality, to make it statistically comparable to well-watered control [138].

Both root development and the rates of K^+ root diffusion were reduced during drought stress, reducing the uptake of K. Drought resistance as well as K absorption might be further depressed by the ensuing reduced levels of K. Consequently, maintenance of sufficient plant K is important for the dryness of plants. A strong relationship has been established between K's nutritional status and plant drought resistance. Adequate quantities of K can increase the overall dry mass accumulation under drought stress compared to lower K levels. The stomata control by K^+ and the associated increased rates of photosynthesis may be related to this result. Furthermore, K is also crucial to the transfer of root development photo-assimilates. Root growth promotion was shown to enhance the root surface that was exposed to the soil as a consequence of improved absorption of root water by increasing adequate K supply under K-deficient soil. Lindhauer said that K nutrition not only enhanced the total dry mass of the plant and leaf area but also improved the retention of water in the drought of plant tissue. Drought stress has considerably decreased cell membrane stability [139]. Maize plants with a higher application of K were shown to respond more effectively to water stress in research [140]. These improvements were mostly due to K's function in enhancing stability and adjusting the capabilities of the cell membrane. A suitable supply of K is necessary to increase dryness by increasing root length and keeping the cell membrane stable.

Many studies have demonstrated that osmotic adjustment is positively linked to drought resistance in different plant types [141]. K^+ plays a crucial function in developing the capacity to adapt to drought environments as one of the most prominent inorganic osmotic in plants [142]. Increasing the K^+, Cl^-, and Na^+ absorption by root cells regulated cell turgor restoration under osmotically-generated stress, partially mediated by K+ transporters on the cellular plasma membrane [143]. Moreover, enough K promotes solvent accumulation, therefore decreasing the osmotic potential to sustain osmotic stress in plant cell turgor. In a word, an adequate K status can enable osmotic adjustments which maintain high turgor pressure, the relative water content, and reduced osmotic potential [144].

For seed adjustment to drought environments, rapid stoma closure and internal moisture conservation are critical. K plays a vital function during stomata activity in turgor control in guard cells [142]. Considering that stomata closure is preceded by the quick release of K+ from the protective cells into the leaf apoplast, it is fair to believe that stomata could hardly be left open in K-deficient circumstances. Some studies show that K shortage in various agricultural plants might cause stomata closures and decrease photosynthesis rates [98, 145]. Rapid stomatal closure and internal humidity conservation are important for seed adaptability to drought conditions. K plays a key role in the turgor regulation activities in guard cells [142]. Since the closure of the stomata is before the rapid release of K+ in leaf apoplast from the protecting cells, it is reasonable that under the K-deficiency situations, stomata could scarcely be left open. Some researchers have shown that K deficit may affect stomata closures and photosynthetic rates in different agricultural plants.

6. Selection and variation for K-uptake in cotton germplasm

In the area of genetic resource management, it is vital for sustainable cotton production to identify the genotypes of cotton for low K input. Production costs and management of K resources in agro environments would be reduced. Cultivation of nutrient-efficient crop genotypes is an effort to improve the efficiency of fertilizer usage [37] and decrease input costs and nutrient waste [17].

Categorization of cotton cultivars based on their growth performance under nutrient-deficient conditions is essential for the development of K-efficient cultivars in any crop. Soil worldwide is exhausting for the supply of adequate potassium (K) nutrition due to intensive crop cultivation systems. Exploiting genetic variability underlying an efficient K transport system is a viable, cost-effective strategy to increase cotton productivity in low-input production systems [12] stated during his experimentation to characterize 46 diverse cotton cultivars for enhanced K acquisition and utilization efficiency at low (0.26 mM K) and adequate (3.33 mM K) K levels in a sand culture experiment. There exists genetic variation in cotton cultivars for K acquisition and utilization. The Indicators such as DMYI and KUE, based on the mean and standard deviation, can be reliably used for the classification of indigenous cotton germplasm. The cotton cultivars including MNH-886, CYTO-124, FH-142, CIM-554, CIM-707, and IUB-2013 were found to be highly K-efficient and responsive. These cotton cultivars have a great potential for wider adaptation under both low and high-K input agriculture systems and, therefore, may be recommended for cultivation in soils.

Supply of adequate plant growth K improved significantly by 21–50% the growth characteristics of cotton and the yield of seed cotton from five cotton genotypes being observed, including symposia branch number (21%), leaf number (34%), dry biomass with leaves (30%), dry biomass shoot (31%), boll number (50%) and cotton yield (92%). In four cotton cultivars for biomass production and partitioning between various organisms under the K fertilization impact, [133] observed significant differences [81] also reported the differential reaction of two K^+ conventional cotton cultivars to two Bt and discovered that, because of decreased K^+ absorption, the Bt-transgenic cotton cultivars were more vulnerable to the deficit than conventional cultivars. However, it is very preferable to select a wide variety of cotton genotypes to categorize their differential K absorption and use efficiency. Such grading might provide significant basic information to support cotton breeding efforts that achieve high efficiency of K-use.

The variations in the absorption and use of nitrogen [146] and potassium among cotton cultivars were also identified [147]. One of the major techniques for the sustainable strengthening of farming systems revealed by Ref. [148] is to use genotypic variations by selecting and identifying crop genotypes best adapted to unflavored soils using nutrients. In their growth response, K-uptake and efficiency associated substantially with shot dry weight were significantly different, indicating an essential function for the creation of biomass. Higher biomass accumulated in K levels in efficient cultivars in the absorption of K and efficiency. Because K is extremely mobile in plants, genotypic variations in K absorption were linked to alterations to K translocation across cells and the whole plant [37]. Bt-transgenic cotton cultivars tend to have an increasing interest in K fertilizers with the rising usage of transgenic cotton as documented by Ref. [149] as a result of a more sensitive contemporary K shortage than traditional cultivars [149]. Reported that high-yielding, early maturing, and determinant cotton cultivars are more sensitive to drought stress compared with indeterminate cultivars that mature later with small boll numbers and weight.

K-uptake efficient genotypes have special physiological mechanisms to achieve enough K-uptake and efficient genotypes can have a larger area for contact between the root and the earth with a greater root surface absorptive capacity to maintain the soil and root spread gradient according to [37] K-uptake efficient genotypes. In eight cotton cultivars under controlled conditions both in growing chamber and field environments, the genetic diversity in K absorption and application was investigated, that [69]. K-efficient cultivars outperformed the K-inefficient cultivars by 29 and 234%, respectively, for K (dry mass supplied by unit K fixation) productivity (dry mass per unit K accumulates) and K. Under field circumstances with soil-compatible K-deficiency, the K-efficient cultivars generated a 59% greater potential yield (dry weight for each regenerative organ). During blooming, boll development, and seed production stage, however, signs of K insufficiency appeared.

More efficient cultivars are expected to have a favorable influence on the environment in their use and use of soil nutrients because farmers may decrease their chemical use in agriculture and demand significant yields, where fertilizer efficiency can be constrained by chemical and biological responses, drying of topsoil, subsurface limitations, and/or the involvement of disease [37].

7. K-efficient cotton cultivars' relation with water use efficiency

Only limited research work has been done on water use efficiency by using K-efficient cultivars. Our findings suggest that the selection of K-efficient cotton cultivars with adequate K application has the potential to improve nitrogen-use efficiency and save considerable quantities of fertilizer and irrigation water, up to 35% in cotton, without yield loss under arid climatic conditions [11]. The authors [135], in their experiment, describe the degree of association between K levels and physiological processes implying that a high supply of K must be aimed at under dry land conditions for improvement in drought tolerance or WUE should be explored genetically despite adopting optimum mineral nutrition and water management strategies. The normal dose of potassium fertilizer reduces the impact of stress and improves water use efficiency by up to 30% by reducing leaf evapotranspiration and preventing water losses, and maintaining leaf water potential [11].

8. Conclusion

The world population will increase up to 9 billion in the year 2050. The biotic and ambitious factors are the main cause of the reduction of yield as well as quality. Therefore, it is necessary to reduce the massive application of fertilizers; however, overuse of N fertilization and K deficiency become a source of poor yield and quality. Recently, more attention has been focused on the effects of morpho-physiological traits that ultimately determine yield. Although all stages of cotton development are affected by drought stress, however, the reproductive phase of flowering and boll development is generally accepted as the most sensitive stages. Therefore, it is necessary to develop nutrient-efficient cultivars to fulfill the yield gap in many regions. According to our knowledge, little information is available for the characterization of cotton cultivars for potassium uptake under drought stress. The knowledge of drought tolerance of K-efficient cotton cultivars during drought stress is crucial for maintaining yield production in regions where water supply is limited or annual rainfall is less.

Acknowledgements

The authors would like to thank the Chief Scientist of Pesticide Quality Control Laboratory, Multan- Pakistan for financial support and writing up.

Author contributions

All authors read, revised, and approved the manuscript.

Funding

Not applicable.

Availability of data and materials

The datasets used and/or analyzed during the current study are available from the corresponding author upon reasonable request.

Declarations

Ethics approval and consent to participate.
Not applicable.

Consent for publication

All authors have reviewed the manuscript and given their consent for publication.

Competing interests

Not applicable.

Author details

Muhammad Naeem Akhtar[1*], Muhammad Waseem Akhtar[2], Ashfaq Ahmad Rahi[1] and Tanveer ul Haq[3]

1 Pesticide Quality Control Laboratory, Multan, Pakistan

2 Department of Plant Breeding and Genetic, University of Agriculture Faisalabad, Pakistan

3 Department of Soil and Environmental Sciences, MNS- University of Agriculture, Multan, Pakistan

*Address all correspondence to: muhammaduam01@gmail.com

References

[1] Riemann M, Dhakarey R, Hazman M, Miro B, Kohli A, Nick P. Exploring jasmonates in the hormonal network of drought and salinity responses. Frontiers in Plant Science. 2015;**6**:1077

[2] Sun Z, Wang X, Liu Z, Gu Q, Zhang Y, Li Z, et al. Genome-wide association study discovered genetic variation and candidate genes of fibre quality traits in Gossypium hirsutum L. Plant Biotechnology Journal. 2017;**15**:982-996

[3] Pervez H, Ashraf M, Makhdum MI. Influence of potassium nutrition on gas exchange characteristics and water relations in cotton (Gossypium hirsutum L.). Photosynthetica. 2004;**42**(2):251-255

[4] Uzilday B, Turkan I, Sekmen AH, Ozgur RENG, Karakaya HC. Comparison of ROS formation and antioxidant enzymes in Cleome gynandra (C4) and Cleome spinosa (C3) under drought stress. Plant Science. 2012;**182**:59-70

[5] Dawn News. 2016. Available from: http://www.dawn.com/news/1240448

[6] Comas LH, Becker SR, Cruz VMV, Byrne PF, Dierig DA. Root traits contributing to plant productivity under drought. Frontiers in Plant Science. 2013;**4**:442

[7] Sahito A, Baloch ZA, Mahar A, Otho SA, Kalhoro SA, Ali A, et al. Effect of water stress on the growth and yield of cotton crop (Gossypium hirsutum L.). American Journal of Plant Sciences. 2015;**6**:1027-1039. DOI: 10.4236/ajps.2015.67108

[8] Wang Y, Wu WH. Regulation of potassium transport and signaling in plants. Current Opinion in Plant Biology. 2017;**39**:123-128

[9] Wang X, Mohamed I, Xia Y, Chen F. Effects of water and potassium stresses on potassium utilization efficiency of two cotton genotypes. International Journal of Molecular Sciences. 2014;**14**:833-844

[10] Wang Y, Wu WH. Potassium transport and signaling in higher plants. Annual Review of Plant Biology. 2013;**64**:451-476

[11] Akhtar MN, Ul-Haq T, Ahmad F, Imran M, Ahmed W, Ghaffar A, et al. Application of potassium along with nitrogen under varied moisture regimes improves performance and nitrogen-use efficiency of high- and low-potassium efficiency cotton cultivars. Agronomy. 2022;**2022**(12):502. DOI: 10.3390/agronomy12020502

[12] Akhtar MN, Ul-Haq T, Ahmad F. Evaluation of the response of indigenous cotton cultivars to low potassium stress in hydroponics system. Pakistan Journal of Botany. 2022;**54**(5):1663-1673. DOI: 10.30848/PJB2022-5(4)

[13] Akhtar MN, Ul-Haq T, Ahmad F, Ahmed W, Ghaffar A. Characterization of diverse cotton cultivars for potassium acquisition based on morphological and physiological traits at early growth stage. Pakistan Journal of Botany. 2023;**55**(2). DOI: 10.30848/PJB2023-2(21)

[14] Anjum SA, Xie XY, Wang LC, Saleem MF, Man C, Lei W. Morphological, physiological and biochemical responses of plants to drought stress. African Journal of Agricultural Research. 2011;**6**(9):2026-2032

[15] Ashraf J, Zuo D, Wang Q, Malik W, Zhang Y, Abid MA, et al. Recent insights into cotton functional genomics: Progress and future perspectives.

Plant Biotechnology Journal. 2018;**16**(3):699-713

[16] Bajji M, Kinet JM, Lutts S. The use of the electrolyte leakage method for assessing cell membrane stability as a water stress tolerance test in durum wheat. Plant Growth Regulation. 2002;**36**(1):61-70

[17] Baligar VC, Fageria NK, He ZI. Nutrient use efficiency in plants. Communications in Soil Science and Plant Analysis. 2001;**32**:921-950

[18] Bandurska H, Stroiński A, Kubiś J. The effect of jasmonic acid on the accumulation of ABA, proline and spermidine and its influence on membrane injury under water deficit in two barley genotypes. Acta Physiologiae Plantarum. 2003;**25**(3):279-285

[19] Baytar AA, Peynircioğlu C, Sezener V, Basal H, Frary A, Frary A, et al. Identification of stable QTLs for fiber quality and plant structure in upland cotton (G. hirsutum L.) under drought stress. Industrial Crops and Products. 2018;**124**:776-786

[20] Bengough AG, McKenzie BM, Hallett PD, Valentine TA. Root elongation, water stress, and mechanical impedance: A review of limiting stresses and beneficial root tip traits. Journal of Experimental Botany. 2011;**62**:59-68

[21] Borgo L, Marur CJ, Vieira LGE. Effects of high proline accumulation on chloroplast and mitochondrial ultrastructure and on osmotic adjustment in tobacco plants. Acta Scientiarum Agronomy. 2015;**37**:191

[22] Boudsocq M, Lauriere C. Osmotic signaling in plants. Multiple pathways mediated by emerging kinase families. Plant Physiology. 2005;**138**(3):1185-1194

[23] Brar MS, Tiwari KN. Boosting seed cotton yields in Punjab with potassium: A review. Better Crops. 2004;**88**(3):28-31

[24] Bray EA. Abscisic acid regulation of gene expression during water-deficit stress in the era of the Arabidopsis genome. Plant, Cell & Environment. 2002;**25**(2):153-161

[25] Brill E, van Thournout M, White RG, Llewellyn D, Campbell PM, Engelen S, et al. A novel isoform of sucrose synthase is targeted to the cell wall during secondary cell wall synthesis in cotton fiber. Plant Physiology. 2011;**157**:40-54

[26] Chan KX, Wirtz M, Phua SY, Estavillo GM, Pogson BJ. Balancing metabolites in drought: The sulfur assimilation conundrum. Trends in Plant Science. 2013;**18**(1):18-29

[27] Chastain DR, Snider JL, Choinski JS, Collins GD, Perry CD. Leaf ontogeny strongly influences photosynthetic tolerance to drought and high temperature in Gossypium hirsutum. Journal of Plant Physiology. 2016;**199**:8-28

[28] Chastain DR, Snider JL, Collins GD, Perry CD, Whitaker J, Byrd SA. Water deficit in field-grown Gossypium hirsutum primarily limits net photosynthesis by decreasing stomatal conductance, increasing photorespiration, and increasing the ratio of dark respiration to gross photosynthesis. Journal of Plant Physiology. 2014;**171**(17):1576-1585

[29] Chaves MM, Flexas J, Pinheiro C. Photosynthesis under drought and salt stress: Regulation mechanisms from whole plant to cell. Annals of Botany. 2009;**103**:551-560

[30] Chen T, Li W, Hu X, Guo J, Liu A, Zhang B. A cotton MYB transcription factor, GbMYB5, is positively involved in plant adaptive response to drought stress. Plant and Cell Physiology. 2015;**56**(5):917-929

[31] Chen TH, Murata N. Glycinebetaine protects plants against abiotic stress: Mechanisms and biotechnological applications. Plant, Cell & Environment. 2011;**34**(1):1-20

[32] Chini A, Fonseca S, Fernandez G, Adie B, Chico JM, Lorenzo O, et al. The JAZ family of repressors is the missing link in jasmonate signalling. Nature. 2007;**448**:666-671

[33] Jamshed M, Jia F, Gong J, Palanga KK, Shi Y, Li J, et al. Identification of stable quantitative trait loci (QTLs) for fiber quality traits across multiple environments in Gossypium hirsutum recombinant inbred line population. BMC Genomics. 2016;**17**:197

[34] FAO. The State of Food and Agriculture. Vol. 37. Rome, Italy: Food & Agriculture Organization of the UN (FAO); 1989

[35] Parekh MJ, Kumar S, Zala HN, Fougat RS, Patel CB, Bosamia TC, et al. Development and validation of novel fiber relevant dbEST-SSR markers and their utility in revealing genetic diversity in diploid cotton (Gossypium herbaceum and G. arboreum). Industrial Crops and Products. 2016;**83**:620-629

[36] Samarah NH. Effects of drought stress on growth and yield of barley. Agronomy for Sustainable Development. 2005;**25**(1):145-149

[37] Riaz M, Farooq J, Sakhawat G, Mahmood A, Sadiq MA, Yaseen M. Genotypic variability for root/shoot parameters under water stress in some advanced lines of cotton (Gossypium hirsutum L.). Genetics and Molecular Research. 2013;**12**(1):552-561

[38] Akhtar ME, Aneela S, Ashraf M, Akhtar M, Zameer Khan M. Effect of potash application on seed cotton yield and yield components of selected cotton varieties-I. Asian Journal of Plant Sciences. 2003;**2**:602-604

[39] Iqbal M, Khan MA, Naeem M, Aziz U, Afzal J, Latif M. Inducing drought tolerance in upland cotton (Gossypium hirsutum L.), accomplishments and future prospects. World Applied Sciences Journal. 2013;**21**:1062-1069

[40] Wang Z, Huang B. Physiological recovery of Kentucky bluegrass from simultaneous drought and heat stress. Crop Science. 2004;**44**(5):1729-1736

[41] Hake K, Cassman K, Ebelhar W. Cotton nutrition-N, P and K. Cotton Physiology Today. 1991;**2**:3

[42] Zia-Ul-Hassan, K.S. Memon, M. Memon and M. Arshad. 2008. Quantifying the effect of temperature on ammonium bicarbonate diethylene triamine penta-acetic acid extractable potassium and developing a novel correction factor to express the data. Communications in Soil Science and Plant Analysis 39:3047-3056.

[43] Farooq M, Wahid A, Kobayashi N, Fujita SMA, Basra D. Plant drought stress: Effects, mechanisms and management. Agronomy for Sustainable Development. 2009;**29**;185-212

[44] Farooq M, Wahid A, Lee DJ. Exogenously applied polyamines increase drought tolerance of rice by improving leaf water status, photosynthesis and membrane properties. Acta Physiologiae Plantarum. 2009;**31**(5):937-945

[45] Loka DM, Derrick M, Oosterhuis DM, Ritchie GL. Water-deficit stress in cotton. In: Oosterhuis DM, editor. Stress Physiology in Cotton. Tennessee, USA: National Cotton Council of America; 2011. pp. 37-72. Number Seven The Cotton Foundation Book Series

[46] Gunes A, Kadioglu YK, Pilbeam DJ, Inal A, Coban S, Aksu A. Influence of silicon on sunflower cultivars under drought stress, II: Essential and nonessential element uptake determined by polarized energy dispersive X-ray fluorescence. Communications In Soil Science and Plant Analysis. 2008;**39**:1904-1927

[47] Kuromori T, Miyaji T, Yabuuchi H, Shimizu H, Sugimoto E, Kamiya A, et al. ABC transporter AtABCG25 is involved in abscisic acid transport and responses. Proceedings of the National Academy of Sciences. 2010;**107**(5):2361-2366

[48] Tan J, Tu L, Deng F, Wu R, Zhang X. Exogenous jasmonic acid inhibits cotton fiber elongation. Journal of Plant Growth Regulation. 2012;**31**(4):599-605

[49] Nam NH, Chauhan YS, Johansen C. Effect of timing of drought stress on growth and grain yield of extra-short-duration pigeonpea lines. The Journal of Agricultural Science. 2001;**136**(2):179-189

[50] Loka DA, Oosterhuis DM, Ritchie GL. Water-deficit stress in cotton. Stress Physiol. Cott. Found. Memphis, TN In D.M Oos. 2011. pp. 37-72

[51] Makhdum MI, Pervez H, Ashraf M. Dry matter accumulation and partitioning in cotton (Gossypium hirsutum L.) as influenced by potassium fertilization. Biology and Fertility of Soils. 2007;**43**(3):295-301

[52] Ray JD, Sinclair TR. Stomatal closure of maize hybrids in response to drying soil. Crop Science. 1997;**37**:803-807

[53] Deeba F, Pandey AK, Ranjan S, Mishra A, Singh R, Sharma YK, et al. Physiological and proteomic responses of cotton (Gossypium herbaceum L.) to drought stress. Plant Physiology and Biochemistry. 2012;**53**:6-18

[54] Mahan JR, Wanjura DF. Seasonal patterns of glutathione and ascorbate metabolism in field-grown cotton under water stress. Crop Science. 2005;**45**:193-201

[55] Marschner P. Marschner's Mineral Nutritionmineral Nutrition of Higher Plants. London: Academic; 2012. p. 651

[56] Estrada-Campuzano G, Miralles DJ, Slafer GA. Genotypic variability and response to water stress of pre- and post anthesis phases in triticale. European Journal of Agronomy. 2008;**28**:171-177

[57] Yamaguchi-Shinozaki K, Shinozaki K. Transcriptional regulatory networks in cellular responses and tolerance to dehydration and cold stresses. Annual Review of Plant Biology. 2006;**57**:781-803

[58] Flexas J, Bota J, Cifre J, Mariano Escalona J, Galmés J, Gulías J, et al. Understanding down-regulation of photosynthesis under water stress: Future prospects and searching for physiological tools for irrigation management. Annals of Applied Biology. 2004;**144**(3):273-283

[59] Saeed M, Guo W, Ullah I, Tabbasam N, Zafar Y, ur-Rahman M, et al. QTL mappin for physiology, yield and plant architecture traits in cotton (Gossypium hirsutum L.) grown under well-watered versus water-stress conditions. Electronic Journal of Biotechnology. 2011;**14**:1-13

[60] Pettigrew WT, Heitholt JJ, Meredith WR Jr. Genotypic interactions with potassium and nitrogen in cotton of varied maturity. Agronomy Journal. 1996;**88**(1):89-93

[61] Seo M, Koshiba T. Complex regulation of ABA biosynthesis in plants. Trends in Plant Science. 2002;**7**(1):41-48

[62] Noctor G, Gomez L, Vanacker H, Foyer CH. Interactions

between biosynthesis, compartmentation and transport in the control of glutathione homeostasis and signalling. Journal of Experimental Botany. 2002;**53**(372):1283-1304

[63] Khan MB, Hussain M, Raza A, Farooq S, Jabran K. Seed priming with CaCl2 and ridge planting for improved drought resistance in maize. Turkish Journal of Agriculture and Forestry. 2015;**39**:193-203

[64] Maréchaux I, Bartlett MK, Sack L, Baraloto C, Engel J, Joetzjer E, et al. Drought tolerance as predicted by leaf water potential at turgor loss point varies strongly across species within an Amazonian forest. Functional Ecology. 2015;**29**:1268-1277

[65] Osakabe Y, Osakabe K, Shinozaki K, Tran L-SP. Response of plants to water stress. Frontiers in Plant Science. 2014;**5**:1-8

[66] Zhang H, Shen G, Kuppu S, Gaxiola R, Payton P. Creating drought- and salt-tolerant cotton by overexpressing a vacuolar pyrophosphatase gene. Plant Signaling & Behavior. 2011;**6**:861-863

[67] Taiz L, Zeiger E. Plant Physiology. 5th ed. Massachusetts: Sinauer Assoc.; 2010. p. 782

[68] Noreen S, Athar HUR, Ashraf M. Interactive effects of watering regimes and exogenously applied osmoprotectants on earliness indices and leaf area index in cotton (Gossypium hirsutum L.) crop. Pakistan Journal of Botany. 2013;**45**(6):1873-1881

[69] Zhao D, Oosterhuis DM, Bednarz CW. Influence of potassium deficiency on photosynthesis, chlorophyll content, and chloroplast ultrastructure of cotton plants. Photosynthetica. 2001;**39**(1):103-109

[70] Kim HJ, Tang Y, Moon HS, Delhom CD, Fang DD. Functional analyses of cotton (Gossypium hirsutum L.) immature fiber (im) mutant infer that fiber cell wall development is associated with stress responses. BMC Genomics. 2013;**14**:889

[71] Kaya MD, Okçu G, Atak M, Cıkılı Y, Kolsarıcı Ö. Seed treatments to overcome salt and drought stress during germination in sunflower (Helianthus annuus L.). European Journal of Agronomy. 2006;**24**(4):291-295

[72] USDA. 2015. Available from: http://www.ers.usda.gov/dataproducts/chartgallery/detail.aspx?chartId=52780

[73] Abbate PE, Dardanellib JL, Cantareroc MG, Maturanoc M, Melchiorid RIM, Sueroa EE. Climatic and water availability effects on water-use efficiency in wheat. Crop Science. 2004;**44**:474-483

[74] Costa LD, Vedove GD, Gianquinto G, Giovanardi R, Peressotti A. Yield, water use efficiency and nitrogen uptake in potato: Influence of drought stress. Potato Research. 1997;**40**:19-34

[75] Fang Y, Xiong L. General mechanisms of drought response and their application in drought resistance improvement in plants. Cellular and Molecular Life Sciences. 2015;**72**(4):673-689

[76] Lv S, Yang A, Zhang K, Wang L, Zhang J. Increase of glycinebetaine synthesis improves drought tolerance in cotton. Molecular Breeding. 2007;**20**:233-248

[77] Manikavelu A, Nadarajan N, Ganesh SK, Gnanamalar RP, Babu RC. Drought tolerance in rice: Morphological and molecular genetic consideration. Plant Growth Regulation. 2006;**50**:121-138

[78] Zhang H, Zhang G, Lü X, Zhou D, Han X. Salt tolerance during seed germination and early seedling stages of 12 halophytes. Plant and Soil. 2015;**388**(1):229-241

[79] Luo J, Zhao LL, Gong SY, Sun X, Li P, Qin LX, et al. A cotton mitogen-activated protein kinase (GhMPK6) is involved in ABA-induced CAT1 expression and H_2O_2 production. Journal of Genetics and Genomics. 2011;**38**:557-565

[80] Wang M, Zheng Q, Shen Q, Guo S. The critical role of potassium in plant stress response. International Journal of Molecular Sciences. 2013;**14**(4):7370-7390

[81] Lu YX, Chanroj S, Zulkifli L, Johnson MA, Uozumi N, Cheung A, et al. Pollen tubes lacking a pair of K^+ transporters fail to target ovules in Arabidopsis. The Plant Cell. 2011;**23**:81-93

[82] Gibbs J, Greenway H. Mechanisms of anoxia tolerance in plants. I. Growth, survival and anaerobic catabolism. Functional Plant Biology. 2003;**30**(1):1-47

[83] Loka DA, Oosterhuis DM. Effect of 1-MCP on gas exchange and carbohydrate concentrations of the cotton flower and subtending leaf under water-deficit stress. American Journal of Plant Sciences. 2013;**04**:142-152

[84] Zahoor R, Zhao W, Abid M, Dong H, Zhou Z. Potassium application regulates nitrogen metabolism and osmotic adjustment in cotton (Gossypium hirsutum L.) functional leaf under drought stress. Journal of Plant Physiology. 2017;**215**:30-38

[85] Ahoor R, Zhao W, Abid M, Dong H, Zhou Z. Potassium application regulates nitrogen metabolism and osmotic adjustment in cotton (Gossypium hirsutum L.) functional leaf under drought stress. Journal of Plant Physiology. 2017;**215**:3025

[86] Galmés J, Flexas J, Savé R, Medrano H. Water relations and stomatal characteristics of Mediterranean plants with different growth forms and leaf habits: Responses to water stress and recovery. Plant and Soil. 2007;**290**:139-155

[87] Abid M, Tian Z, Ataulkarim ST, Cui Y, Liu Y. Nitrogen nutrition improves the potential of wheat (Triticum aestivum L.) to alleviate the effects of drought stress during vegetative growth periods. Frontiers in Plant Science. 2016;7:1

[88] Hetherington AM, Woodward FI. The role of stomata in sensing and driving environmental change. Nature. 2003;**424**(6951):901-908

[89] Wu J, Sun Y, Zhao Y, Zhang J, Luo L, Li M, et al. Deficient plastidic fatty acid synthesis triggers cell death by modulating mitochondrial reactive oxygen species. Cell Research. 2015;**25**(5):621-633

[90] Singh P. Cotton Breeding. Cuttack, Ludhiana New Delhi Noida (U.P.): Kalyani Publishers; 2004. p. 295

[91] Girma K, Teal RK, Freeman KW, Boman RK, Raun WR. Cotton lint yield and quality as affected by applications of N, P, and K fertilizers. J Cotton Sci. 2007;**11**:12-19

[92] Singh M, Kumar J, Singh S, Singh VP, Prasad SM. Roles of osmoprotectants in improving salinity and drought tolerance in plants: A review. Reviews in Environmental Science and Bio/ Technology. 2015;**14**(3):407-426

[93] Shang L, Wang Y, Wang X, Liu F, Abduweli A, Cai S. Genetic analysis

and QTL detection on fiber traits using two recombinant inbred lines and their backcross populations in upland cotton. Genes - Genomes - Genetics. 2016;**6**(9):2717-2724

[94] Parida AK, Dagaonkar VS, Phalak MS, Umalkar G, Aurangabadkar LP. Alterations in photosynthetic pigments, protein and osmotic components in cotton genotypes subjected to short-term drought stress followed by recovery. Plant Biotechnology Reports. 2007;**1**:37-48

[95] Jin SH, Huang JQ, Li XQ, Zheng BS, Wu JS, Wang ZJ, et al. Effects of potassium supply on limitations of photosynthesis by mesophyll diffusion conductance in Carya cathayensis. Tree Physiology. 2011;**31**(10):1142-1151

[96] Shen X, Guo T, Zhu W, Zhang X. Mapping fiber and yield QTLs with main epistatic and QTL environment interaction effects in recombinant inbred lines of upland cotton. Crop Science. 2006;**46**:61-66

[97] Samal D, Kovar JL, Steingrobe B, Sadana US, Bhadoria PS, Claassen N. Potassium uptake efficiency and dynamics in the rhizosphere of maize (Zea mays L.), wheat (Triticum aestivum L.), and sugar beet (Beta vulgaris L.) evaluated with a mechanistic model. Plant and Soil. 2010;**332**:105-121

[98] Kamara AY, Menkir A, Badu-Apraku B, Ibikunle 0. The influence of drought stress on growth, yield and yield components of selected maize genotypes. The Journal of Agricultural Science. 2003;**141**:43-50

[99] Liu G, Li X, Jin S, Liu X, Zhu L, Nie Y, et al. Overexpression of rice NAC gene SNAC1 improves drought and salt tolerance by enhancing root development and reducing transpiration rate in transgenic cotton. PLoS One. 2014;**9**:e86895

[100] Hulugalle NR, Broughton KJ, Tan DKY. Root growth of irrigated summer crops in cotton-based farming systems sown in Vertosols of northern New South Wales. Crop and Pasture Science. 2015;**66**(2):158-167

[101] Yang XE, Liu JX, Wang WM, Li H, Luo AC, Ye ZQ, et al. Genotypic differences and some associated plant traits in potassium internal use efficiency of lowland rice (Oryza sativa L.). Nutrient Cycling in Agroecosystems. 2003;**67**(3):273-282

[102] Deng F, Tu L, Tan J, Li Y, Nie Y, Zhang X. GbPDF1 is involved in cotton fiber initiation via the core cis-element HDZIP2ATATHB2. Plant Physiology. 2012;**158**:890-904

[103] Rengel Z, Damon PM. Crops and genotypes differ in efficiency of potassium uptake and use. Physiologia Plantarum. 2008;**133**(4):624-636

[104] Rucker KS, Kvien CK, Holbrook CC, Hook JE. Identification of peanut genotypes with improved drought avoidance traits. Peanut Science. 1995;**22**(1):14-18

[105] Xu Z, Zhou G, Shimizu H. Plant responses to drought and rewatering. Plant Signaling & Behavior. 2010;**5**:649-654

[106] Kitao M, Lei TT. Circumvention of over-excitation of PSII by maintaining electron transport rate in leaves of four cotton genotypes developed under long-term drought. Plant Biology. 2007;**9**:69-76

[107] Mafakheri A, Siosemardeh A, Bahramnejad B, Struik PC, Sohrabi E. Effect of drought stress on yield, proline and chlorophyll contents in three chickpea cultivars. Australian Journal of Crop Science. 2010;**4**:580-585

[108] Zhang H, Khan A, Tan DK, Luo H. Rational water and nitrogen management improves root growth, increases yield and maintains water use efficiency of cotton under mulch drip irrigation. Frontiers in Plant Science. 2017;**8**:912

[109] Wan J, Rebecca G, Ying J, Peter MC, Huang Y. Development of drought tolerant canola (Brassica napus L.) through genetic modulation of ABA-mediated stomatal responses. Crop Science. 2009;**49**:1539-1554

[110] Yadav RS, Hash CT, Bidinger FR, Devos KM, Howarth CJ. Genomic regions associated with grain yield and aspects of post-flowering drought tolerance in pearl millet across stress environments and tester background. Euphytica. 2004;**136**(3):265-277

[111] Li DX, Li CD, Sun HC, Liu LT, Zhang YJ. Photosynthetic and chlorophyll fluorescence regulation of upland cotton (Gossiypium hirsutum L.) under drought conditions. Plant Omics. 2012;**5**:432-437

[112] Wang C, Lu W, He X, Wang F, Zhou Y, Guo X, et al. The cotton mitogen-activated protein kinase kinase 3 functions in drought tolerance by regulating stomatal responses and root growth. Plant & Cell Physiology. 2016;**57**:1629-1642

[113] Hu H, He X, Tu L, Zhu L, Zhu S, Ge Z, et al. GhJAZ2 negatively regulates cotton fiber initiation by interacting with the R2R3-MYB transcription factor GhMYB25-like. The Plant Journal. 2016;**88**:921-935

[114] Wendel JF, Cronn RC. Polyploidy and the evolutionary history of cotton. Advances in Agronomy. 2002;**78**:139-186

[115] Premachandra GS, Saneoka H, Ogata S. Cell membrane stability and leaf water relations as affected by potassium nutrition of water-stressed maize. Journal of Experimental Botany. 1991;**42**(6):739-745

[116] Statista. 2015. Available from: http://www.statista.com/statistics/263055/cotton-production-worldwide-by-top-countries/

[117] Okcu G, Kaya MD, Atak M. Effects of salt and drought stresses on germination and seedling growth of pea (Pisum sativum L.). Turkish Journal of Agriculture and Forestry. 2005;**29**:237-242

[118] Divya K, Jami SK, Kirti PB. Constitutive expression of mustard annexin, AnnBj1 enhances abiotic stress tolerance and fiber quality in cotton under stress. Plant Molecular Biology. 2010;**73**(3):293-308

[119] Zhang W, Jin FF, Turner A. Increasing autumn drought over southern China associated with ENSO regime shift. Geophysical Research Letters. 2014;**41**(11):4020-4026

[120] Shabala SN, Lew RR. Turgor regulation in osmotically stressed Arabidopsis epidermal root cells. Direct support for the role of inorganic ion uptake as revealed by concurrent flux and cell turgor measurements. Plant Physiology. 2002;**129**(1):290-299

[121] Machado A, Wu Y, Yang Y, Llewellyn DJ, Dennis ES. The MYB transcription factor GhMYB25 regulates early fibre and trichome development. The Plant Journal. 2009;**59**:52-62

[122] Tomemori H, Hamamura K, Tanabe K. Interactive effects of sodium and potassium on the growth and photosynthesis of spinach and komatsuna. Plant Production Science. 2002;**5**(4):281-285

[123] Ruan YL, Llewellyn DJ, Furbank RT. Suppression of sucrose synthase

gene expression represses cotton fiber cell initiation, elongation, and seed development. The Plant Cell. 2003;**15**:952-964

[124] Gill SS, Tuteja N. Reactive oxygen species and antioxidant machinery in abiotic stress tolerance in crop plants. Plant Physiology and Biochemistry. 2010;**48**(12):909-930

[125] Nonaini H. Plant water relations and control of cell elongation at low water potentials. Journal of Plant Research. 1998;**111**:373-382

[126] Das K, Roychoudhury A. Reactive oxygen species (ROS) and response of antioxidants as ROS-scavengers during environmental stress in plants. Frontiers in Environmental Science. 2014;**2**:53

[127] Xu WL, Zhang DJ, Wu YF, Qin LX, Huang GQ, Li J, et al. Cotton PRP5 gene encoding a proline-rich protein is involved in fiber development. Plant Molecular Biology. 2013;**82**:353-365

[128] Von CS, Lawson T, Oxborough K, Baker NR, Andrews TJ, Raines CA. Stomatal conductance does not correlate with photosynthetic capacity in transgenic tobacco with reduced amounts of rubisco. Journal of Experimental Botany. 2004;**55**:1157-1166

[129] Taiz L, Zeiger E. Plant Physiology. 4th ed. Sunderland, MA: Sinauer Associates Inc Publishers; 2006

[130] Zhao TJ, Sun S, Liu Y, Liu IM, Liu Q, Yan YB, et al. Regulating the drought-responsive element (DRE)-mediated signaling pathway by synergic functions of trans-active and transinactiveDRE binding factors in Brassica napus. The Journal of Biological Chemistry. 2006;**281**:10752-10759

[131] Zeid IM, Shedeed ZA. Response of alfalfa to putrescine treatment under drought stress. Biologia Plantarum. 2006;**50**:635-640

[132] Pettigrew WT. Physiological consequences of moisture deficit stress in cotton. Crop Science. 2004;**44**(4):1265-1272

[133] Manavalan LP, Guttikonda SK, Tran LSP, Nguyen HT. Physiological and molecular approaches to improve drought resistance in soybean. Plant & Cell Physiology. 2009;**50**:1260-1276

[134] Zia-ul-Hassan and M. Arshad. Cotton growth under potassium deficiency stress is influenced by photosynthetic apparatus and root system. Pakistan Journal of Botany. 2010;**42**:917-925

[135] Pettigrew WT. Relationships between insufficient potassium and crop maturity in cotton. Agronomy Journal. 2003;**95**:1323-1329

[136] Turner NC, Wright GC, Siddique KHM. Adaptation of grain legumes (pulses) to water-limited environments. Advances in Agronomy. 2001;**71**:193-131

[137] Ahmad R, Hur RGM, Waraich EA, Ashraf MY, Hussain M. Effect of supplemental foliar-applied potassium on cotton (Gossypium hirsutum L.) yield and lint quality under drought stress. Pakistan Journal of Life and Social Sciences. 2013;**11**:154-164

[138] Ahmad MK, Khan AA, Mahmood R. Taurine ameliorates potassium bromate-induced kidney damage in rats. Amino Acids. 2013;**45**(5):1109-1121

[139] Wang Z, Gerstein M, Snyder M. RNA-Seq: A revolutionary tool for transcriptomics. Nature Reviews Genetics. 2009;**10**(1):57-63

[140] Ratnayaka HH, Molin WT, Sterling TM. Physiological and

antioxidant responses of cotton and spurred anoda under interference and mild drought. Journal of Experimental Botany. 2003;**54**:2293-2305

[141] DaCosta M, Huang B. Osmotic adjustment associated with variation in bentgrass tolerance to drought stress. Journal of the American Society for Horticultural Science. 2006;**131**(3):338-344

[142] Mir RR, Zaman-Allah M, SreenivasuluN,TrethowanR,VarshneyRK. Integrated genomics, physiology and breeding approaches for improving drought tolerance in crops. Theoretical and Applied Genetics. 2012;**125**:625-645

[143] Shakoor MS, Malik TA, Azhar FM, Saleem MF. Genetics of agronomic and fiber traits in upland cotton under drought stress. International Journal of Agriculture and Biology. 2010;**12**:495-500

[144] Egilla JN, Davies FT, Boutton TW. Drought stress influences leaf water content, photosynthesis, and water-use efficiency of Hibiscus rosa-sinensis at three potassium concentrations. Photosynthetica. 2005;**43**(1):135-140

[145] Tsonev T, Velikova V, Yildiz-Aktas L, Gürel AYNUR, Edreva A. Effect of water deficit and potassium fertilization on photosynthetic activity in cotton plants. Plant Biosystems-An International Journal Dealing with all Aspects of Plant Biology. 2011;**145**(4):841-847

[146] Pilon C, Oosterhuis DM, Ritchie G, de Paiva Oliveira EA. Effect of drought in the osmotic adjustment of cotton plants. Summaries of Arkansas Cotton Research. 2013:60

[147] Gill MA, Ahmad MI, Yaseen M. Potassium-deficiency stress tolerance and potassium utilization efficiency in wheat genotypes. In: Plant Nutrition for Sustainable Food Production and Environment. Dordrecht: Springer; 1997. pp. 321-322

[148] Yu LH, Wu SJ, Peng YS, Liu RN, Chen X, Zhao P, et al. Arabidopsis EDT1/HDG11 improves drought and salt tolerance in cotton and poplar and increases cotton yield in the field. Plant Biotechnology Journal. 2015;**14**:72-84

[149] Dong H, Tang ZLW, Zhang D. On potassium deficiency in cotton—disorder, cause, and tissue diagnosis. Agriculturae Conspectus Scientificus. 2004;**67**:77-85

Chapter 3

Naturally Colored Cotton: A Sustainable Innovation

Amit Madhu

Abstract

Historically, the global textile industry has been characterized by resource-intensive processes, environmental degradation and a high dependence on synthetic dyes that contribute to pollution. In recent years, the quest for sustainability has led to the exploration of alternative methods and materials that minimize the industry's ecological footprint. Naturally colored cotton fibers, ranging from earthy browns and greens to subdued yellows and reds, are produced by cotton plants that have either been genetically altered or carefully bred. This unique characteristic eliminates the need for chemical dyeing processes, conserving water, energy and chemical usage throughout the textile production cycle. This chapter explores the creative strategy of naturally colored cotton, a potential solution that satisfies both the rising demand for sustainable textiles and environmental concerns. Additionally, discusses possible challenges and issues, such as the agronomic procedures needed to grow cotton with natural colors, the necessity of responsible biotechnology regulation and the integration of these novel fibers into current textile supply chains. It also provides insights into the crucial role that bioengineered fibers can play in defining a more environmentally harmonious and socially responsible future by providing a thorough overview of the advantages, difficulties and prospects of naturally colored cotton.

Keywords: colored cotton, bioengineered fibers, genetic engineering, sustainable fashion, textiles

1. Introduction

The textile industry is undergoing a paradigm shift as sustainability becomes a central concern in every facet of production. Naturally colored cotton has become a popular and environmentally sustainable replacement for commercially dyed cotton during this phase. The pursuit of naturally colored cotton aligns with the growing demand for sustainable practices that reduce the environmental impact of textile production, such as the excessive use of water and chemicals in dyeing processes [1, 2]. This chapter explores the innovative realm of naturally colored cotton, exploring its significance, potential benefits and the diverse techniques employed to cultivate and harness its unique characteristics. From conventional breeding and genetic engineering to mutation breeding and hybridization, the techniques used to produce naturally colored cotton exemplify the intricate interplay between science, agriculture and sustainable fashion. Through this exploration, we aim to shed light on the promising

IntechOpen

prospects and challenges that lie ahead in the realm of naturally colored cotton, underscoring its potential to revolutionize the textile landscape by weaving together vibrant hues with ecological consciousness.

2. Techniques for naturally colored cotton

Producing naturally colored cotton involves breeding and developing cotton varieties that naturally produce fibers with colors other than the typical white or off-white, **Figure 1**. The techniques used to cultivate naturally colored cotton have evolved significantly over the years, primarily driven by advancements in genetics, breeding and agronomy. This can be achieved through conventional breeding techniques or genetic engineering, **Figure 2**. It's important to note that the development of naturally colored cotton involves a combination of these techniques and may require multiple generations to achieve the desired color stability and fiber quality. Additionally, regulatory considerations and consumer acceptance play a role, especially when genetic engineering is involved [3, 4].

2.1 Cross-breeding

Hybridization is a fundamental breeding technique used in the production of colored cotton. It involves controlled cross-breeding between two cotton plants with specific color traits to create offspring with desired color characteristics. By cross-breeding different naturally colored cotton varieties with distinct color traits, breeders can generate hybrids that exhibit a wider range of colors and color intensities in their fibers. Since ancient times, these methods have been employed to create cotton types with certain colors, including brown, green or tints in between [4–6].

The breeding process for naturally colored cotton starts by identifying cotton plants with the desired color traits. This can be done by observing cotton plants in the field and selecting those that naturally exhibit the desired colouration. Now cross-pollinate the selected parent plants with each other, introducing genetic diversity. For example, if you want to develop brown cotton, you would cross-pollinate two brown cotton plants.

Figure 1.
Pictures of naturally colored cotton.

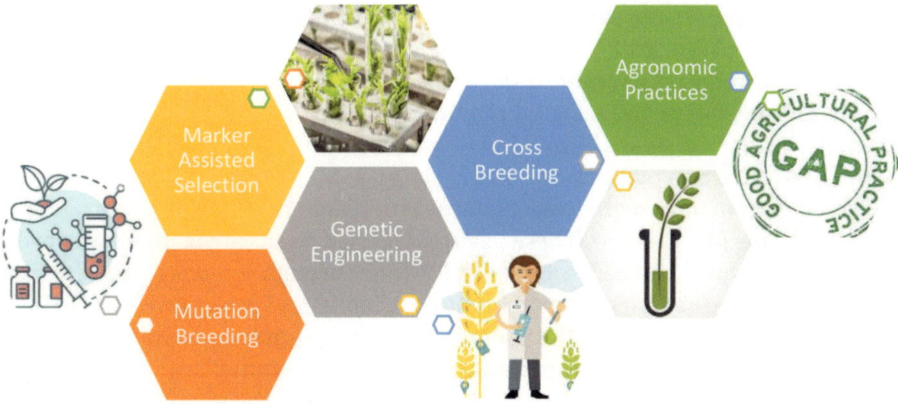

Figure 2.
Techniques for producing naturally colored cotton.

The resulting offspring, known as the first generation or F1 generation, will have a mix of genetic traits from the parent plants. Some may display the desired color, while others may not. Select the plants from the F1 generation that match the desired color trait closely. In the case of brown cotton, select plants with brownish fibers. Now, cross the selected F1 plants with one of the original parent plants that possessed the desired color trait. This is known as backcrossing.

Backcrossing aims to reinforce the genetic traits responsible for the desired color while reducing the influence of other, unwanted genetic traits. This helps in stabilizing the color trait. This backcrossing is repeated for multiple generations until a cotton variety with stable and consistent color traits is achieved. This can take several generations of breeding. Once a stable, naturally colored cotton variety is achieved, scale up production by cultivating these plants on a larger scale for commercial purposes.

Pima Sutra Cotton is a naturally colored cotton variety developed through conventional breeding. It has a range of colors including shades of brown, green and red. Breeders selectively crossed different naturally colored cotton varieties to enhance color expression and fiber quality. These techniques require patience, careful observation and a deep understanding of genetics and plant biology. Over time, these techniques have been successful in producing naturally colored cotton varieties that are both commercially viable and environmentally friendly [6].

2.2 Genetic engineering

Genetic engineering involves introducing specific genes into cotton plants to achieve desired traits, such as natural colouration [3, 5]. This method allows for more precise control over the color traits. Blue Cotton Researchers from CSIRO, an Australian research organization, used genetic engineering to produce blue cotton fibers. They introduced two genes from the pansy flower into cotton plants. These genes encode enzymes responsible for producing blue pigments called anthocyanins. The resulting cotton plants produced fibers with a light blue color [6–8].

Genetic engineering, also known as genetic modification (GM), offers a more targeted and rapid approach to producing naturally colored cotton by introducing specific genes responsible for pigment production into cotton plants. This method involves manipulating the genetic material of the plants to achieve the desired color traits.

In this technique firstly, the genes responsible for pigment production in other naturally colored plants are identified and selected. These genes encode enzymes or proteins involved in the synthesis of pigments like anthocyanins, carotenoids or chlorophyll. Now, isolate the selected pigment-producing genes from their natural sources and insert the isolated genes into a suitable vector, which is typically a small piece of DNA, such as a plasmid or a viral vector [2, 6, 7]. For example, if you want to create green cotton, you might isolate genes responsible for chlorophyll production from green plants like spinach. These vectors are used to carry the pigment-producing genes into the cotton plant cells which can be done through various methods, including Agrobacterium-mediated transformation or biolistics (gene gun).

Once inside the cotton cells, the vector carries the pigment-producing genes into the plant's genome. These genes are integrated into the cotton plant's DNA. Now the transformed cotton cells that have successfully integrated the pigment-producing genes are cultivated to regenerate whole cotton plants with the desired genetic modification.

Genetic engineering techniques offer a more precise and controlled method for producing naturally colored cotton compared to conventional breeding. However, it also involves regulatory and public acceptance considerations due to concerns about GMOs (genetically modified organisms). Careful testing and monitoring are essential to ensure the safety and stability of genetically modified cotton varieties [7, 9].

2.3 Marker-assisted selection (MAS)

Marker-assisted selection (MAS) is a powerful breeding technique used to select plants with specific desirable traits, including natural colouration, based on genetic markers associated with those traits. When it comes to producing naturally colored cotton, MAS can be a valuable tool for breeders to efficiently and accurately identify cotton plants with the desired color characteristics [6, 10].

Initially, researchers identify genetic markers that are linked to the expression of natural color traits in cotton. These markers can be specific DNA sequences or variations (mutations) in certain genes that are consistently associated with the desired color trait. To effectively use MAS, a genetic map of the cotton genome is created. This map helps locate the positions of the genetic markers relative to one another and specific color traits. Then DNA samples are collected from cotton plants, including the ones under consideration for breeding. These samples can be from different cotton varieties or individual plants within the same variety The collected DNA samples are subjected to genetic analysis to identify which plants carry the specific markers associated with the desired color traits. Molecular techniques like PCR (Polymerase Chain Reaction) and DNA sequencing are often used for this purpose [10, 11].

Based on the genetic marker information and trait confirmation, breeders select cotton plants with the most promising genetic profiles for color traits. The selected parent plants are crossbred to create a new generation of cotton plants that carry the desired color traits. This process helps in combining the genetic markers associated with colouration. The offspring from the crossbreeding are then screened using genetic markers to identify individuals who have inherited the desired color traits. This process of selection and breeding is repeated over several generations to stabilize the color traits and improve other characteristics, such as fiber quality and yield [11].

MAS enables breeders to expedite the selection process by focusing on specific genes and markers associated with natural colouration, ultimately leading to the development of naturally colored cotton varieties with enhanced traits. This technique enhances breeding efficiency, reduces the time required to develop new varieties and allows for precise selection based on genetic information [10, 12].

2.4 Mutation breeding

Mutation breeding is a technique used to induce genetic mutations in plants to generate new traits, including natural color variations, that can be advantageous for agriculture or other purposes. In the context of producing naturally colored cotton, mutation breeding can be employed to generate cotton plants with unique color traits through the introduction of genetic mutations [9, 10, 12].

In mutation breeding, cotton seeds or young plants are exposed to mutagenic agents. These agents can include chemicals (e.g., ethyl methane sulfonate), radiation (e.g., gamma rays or X-rays) or even biological mutagens. The mutagenic treatment is carefully controlled to induce mutations without causing excessive damage to the plant's DNA. The goal is to create genetic variation in the cotton population.

After mutagen treatment, the treated seeds are germinated and grown into plants. Each plant represents a potential source of genetic variation. The plants that have undergone mutagenesis are carefully observed for any changes in their phenotypes (observable traits). In the context of natural colouration, breeders are looking for plants that display variations in color compared to the original cotton variety. Plants that exhibit the desired natural color traits are selected as mutants and the selected mutants are propagated through vegetative or generative methods to produce a stable population of cotton plants with the new color trait [4, 12, 13].

Mutation breeding is a powerful tool for introducing novel traits, such as natural colouration, into cotton varieties. It relies on the random creation of genetic variation and the subsequent selection of desirable mutants. While it can produce valuable results, it may also require extensive screening and breeding efforts to stabilize and commercialize the new trait. Additionally, regulatory considerations related to genetically modified organisms (GMOs) may apply in some cases, depending on the extent of the genetic changes induced by the mutagenic treatment [14].

2.5 Agronomic practices

Agronomic practices also play a crucial role in producing naturally colored cotton by influencing the development and preservation of pigments responsible for the colouration of the cotton fibers. Every step, from careful variety selection to specific soil preparation, contributes to the deep and vivid colors that are unique to these cotton cultivars. Growers create a climate that promotes optimum pigment development by strategically fertilizing plants, providing sufficient watering, efficiently controlling pests and diseases and meticulously timing the harvest at peak maturity [2, 4–6].

These procedures protect the color integrity throughout the manufacturing process, together with careful post-harvest processing and quality control methods. These practices, coupled with attentive post-harvest processing and quality control measures, safeguard the integrity of the color throughout the production process. In concert with genetic advancements and environmental considerations, the application of agronomic expertise fosters the growth of naturally colored cotton, yielding fibers that embody the harmonious marriage of sustainable cultivation and captivating colouration [4, 6].

3. Environmental implications

Naturally colored cotton, which is frequently promoted as an environmentally benign substitute for conventionally farmed white cotton, has a number of both advantageous and disadvantageous effects on the environment. These implications stem from both the cultivation and processing of colored cotton, and they often align with sustainability goals. Here are some in-depth remarks on the effects of naturally colored cotton on the environment [2, 6, 14].

Reduced chemical usage: One of the most significant environmental benefits of naturally colored cotton is its potential to reduce the need for synthetic dyes and chemicals. Conventional cotton dyeing procedures use a lot of resources, including a lot of water and chemicals. Naturally colored cotton eliminates the need for these processes, thus lowering water pollution and chemical runoff, which can harm aquatic ecosystems.

Lower carbon footprint: Chemicals and synthetic colors used in conventional cotton dyeing techniques produce a significant amount of carbon emissions. Naturally colored cotton can help reduce greenhouse gas emissions linked to the manufacture and use of dyes by avoiding these procedures.

Water conservation: Since cotton that is naturally colored does not need to be dyed, processing naturally colored cotton frequently uses less water. This can help conserve water resources, which are under increasing pressure in many cotton-growing regions due to irrigation demands.

Biodiversity and ecosystems: Some naturally colored cotton varieties are known to have pest resistance due to their pigment content. This can reduce the need for chemical pesticides, leading to less harm to non-target insects, wildlife and the overall ecosystem.

Soil health: Implementing practices that are conducive to growing naturally colored cotton, such as reduced chemical use and more targeted pest management, can promote healthier soil. Improved water retention, less erosion and greater nutrient cycling are all benefits of healthier soils.

Genetic diversity: Naturally colored cotton varieties are often bred to express specific colors, leading to a broader genetic diversity within cotton cultivation. This diversification can help enhance resilience to pests and diseases, reducing the likelihood of widespread crop failure due to a single pathogen or pest.

While naturally colored cotton presents promising environmental benefits, it also comes with its share of challenges. Sustainable farming methods, such as organic and regenerative agriculture, should be encouraged simultaneously with the production of these distinctive cotton types to maximize the beneficial environmental effects of colored cotton [10, 15, 16].

4. Challenges and future prospects

The integration of naturally colored cotton into the sustainable fashion and textile sector presents both exciting opportunities and complex challenges. While the environmental benefits of reduced chemical usage and lower carbon emissions are evident, challenges unique to the industry must also be addressed. The restricted color selection provided by naturally colored cotton is a significant barrier. Fashion frequently thrives on a wide variety of vibrant hues that may not complement the more subdued tones of naturally colored fibers. This limitation can potentially hinder the creativity of designers seeking vibrant and varied color options for their collections [17, 18].

Moreover, the inconsistent availability of naturally colored cotton in desired quantities poses supply chain challenges. The commercialization of naturally colored cotton hinges on ensuring stable and sufficient production to meet the demands of the fashion industry. However, factors such as varying weather conditions, crop yields and the lack of standardized cultivation practices can lead to fluctuations in supply, potentially disrupting the production schedules of fashion brands. Additionally, processing and manufacturing practices need refinement to cater to the unique characteristics of naturally colored cotton. Developing efficient and cost-effective methods for dyeing, finishing and blending these fibers with other materials is essential. Without appropriate processing solutions, the potential benefits of naturally colored cotton may be overshadowed by higher production costs or compromised fabric quality [16, 18].

Furthermore, consumer preferences often lean towards a wider range of colors than what naturally colored cotton can offer. Raising awareness among consumers about the value and benefits of naturally colored cotton textiles is crucial for market demand to grow. Bridging this gap in perception requires strategic marketing efforts to highlight the unique selling points of these textiles. To address these challenges, ongoing research is necessary to improve the agronomic characteristics of colored cotton varieties, develop efficient processing methods and create greater consumer awareness and demand for these eco-friendly products. As the demand for sustainable and eco-friendly products grows, policymakers and stakeholders need to strike a balance between regulatory oversight and fostering innovation in the cultivation of naturally colored cotton.

The prospects of naturally colored cotton hold promise for sustainable and innovative advancements in the textile industry. As the global awareness of environmental concerns continues to grow, the demand for eco-friendly and ethically produced textiles is on the rise. Naturally colored cotton aligns well with this trend, offering a unique opportunity to reduce the environmental footprint of textile production. Advances in breeding and genetic engineering could lead to the development of naturally colored cotton varieties that not only match the yield and fiber quality of conventional white cotton but also exhibit an even broader range of vibrant and stable colors. Moreover, as research progresses, optimized cultivation techniques and standardized processing methods could make the production of naturally colored cotton more cost-effective and scalable, thereby expanding its market reach [18–20].

Collaborations between agricultural scientists, textile manufacturers and fashion designers could drive innovation and creativity in utilizing these distinctive fibers for a variety of applications, from high-fashion garments to sustainable home textiles. Ultimately, the future of naturally colored cotton holds the potential to reshape the textile industry by offering a harmonious blend of environmental responsibility and artistic expression.

5. Conclusions

Naturally colored cotton represents a sustainable innovation in the textile industry. Techniques such as genetic crossbreeding, molecular genetics and agronomic practices have facilitated its development, offering a promising solution to the environmental challenges associated with conventional cotton dyeing. While challenges exist, ongoing research and increasing consumer awareness may propel the adoption of naturally colored cotton as a viable and sustainable alternative in the textile and fashion industries.

Author details

Amit Madhu
The Technological Institute of Textile and Sciences, Bhiwani, India

*Address all correspondence to: amit.madhu5@gmail.com

References

[1] Chaudhry MR. Natural colours of cotton. ICAC Recorder, Washington. 1992;**10**(4):3-5

[2] Garciai S, de Alencar I. Textile industry can be less pollutant: Introducing naturally coloured cotton. International Journal of Product Management Engineering. 2014;**2**(2):85-91

[3] Hustvedt G, Crews PC. Naturally coloured cotton: Resistance to changes. Journal of Cotton Science. 1999;**3**(1):1-9

[4] Venkatakrishnan S, Mandhyan PK, Kumar K, Narkar R. Studies on the physical properties of naturally coloured cotton and development of eco-friendly UV protected fabric. Technical Textiles. 2015;**5**(6):1-6

[5] Chhabra J, Suri M. Naturally coloured cotton. Asian Textile Journal. 2010;**19**(1):25-29. Available from: https://www.researchgate.net/publication/283581684

[6] Sun J, Sun Y, Zhu QH. Breeding next-generation naturally coloured Cotton. Trends in Plant Science. 2021;**26**(6):539-542

[7] Zhou L, Ye Y. A genetic pathway for naturally coloured cotton production. In Plant Physiology. American Society of Plant Biologists. 2022;**189**(3):1190-1191. DOI: 10.1093/plphys/kiac168

[8] Price JB, Cui X, Calamari TA. Assessing the quality of four naturally coloured Cotton. Textile Research Journal. 2001;**71**(11):993-999

[9] Han A, Chae Y, Lee M. Effect of colour changes of naturally coloured organic Cotton fibres on human sensory perception. Fibres and Polymers. 2011;**12**(7):939-945

[10] Dutt Y, Wang XD, Zhu YG. Breeding for high yield and fibre quality in coloured cotton. Plant Breeding. 2004;**123**:145-151

[11] Cardoso GD, Alves PLCA, Severino LS. Critical periods of weed control in naturally green coloured cotton BRS Verde. Industrial Crops and Products. 2011;**34**:1198-1202

[12] Holton TA, Cornish EC. Genetics and biochemistry of anthocyanin biosynthesis. The Plant Cell. 1995;**7**:1071-1083

[13] Martin C, Prescott A, Mackay S, Bartlett J, Vrijlandt E. Control of anthocyanin biosynthesis in flowers of Antirrhinum majus. The Plant Journal. 1991;**1**:37-49

[14] Wang N, Zhang B, Yao T, Shen C, Wen T, Zhang R, et al. Re enhances anthocyanin and proanthocyanidin accumulation to produce red foliated cotton and brown fibre. Plant Physiology. 2022;**189**:1466-1481

[15] Wen T et al. Linkage and association mapping reveals the genetic basis of brown fibre (Gossypium hirsutum). Plant Biotechnology Journal. 2018;**16**:1654-1666

[16] Peng Z et al. Flavonoid biosynthetic and starch and sucrose metabolic pathways are involved in the pigmentation of naturally brown-coloured cotton fibres. Industrial Crops and Products. 2020;**158**:113045

[17] Khadi BM, Kulkarni BN. Coloured cotton: Problems and their

prospects, the. Indian Textile Journal. 1996;**106**(11):72-76

[18] Rathinamoorthy R, Parthiban M, Cotton C. Novel eco-friendly textile material for the future. In: Martínez L, Kharissova O, Kharisov B, editors. Handbook of Ecomaterials. Cham: Springer; 2019. DOI: 10.1007/978-3-319-68255-6_91

[19] Lolita JC. Colour Cotton Research: Current and Future Perspectives. 2021. Available from: https://www.ijcrt.org

[20] Pavan M, Paschapur NS. Naturally coloured cotton: A game-changer in the textile industry. Just Agriculture. 2022;**2**(8):1-4. Available from: https://www.justagriculture.in

Section 2

Pest Management Practices in Cotton Production

Chapter 4

Role of Biotic and Abiotic Factors for Sustainable Cotton Production

Mehboob Hussain, Xi Gao, Deqiang Qin, Xiaoping Qin and Guoxing Wu

Abstract

Climate is changing globally nowadays because of extensive crucial human activities. This state along with stark in weather measures ultimately affecting the development and growth of crops due to various kinds of stressful field condition at the same time including biotic and abiotic stresses. Thus, various biotic factors including pathogens, weeds and pests and abiotic factors including temperature, humidity and drought etc. are involved in reduction of cotton yield due to which cotton production significantly reduced. Various biotic factors have direct effect on the cotton production and caused significant reduction in cotton crop yield estimated up to 10 to 30%, while as abiotic factors are even worse than biotic stresses and could cause 50% reduction. So, effective agronomic practices, optimal climate and integrated pest management leads to fruitful crop production to cover this yield gap. This chapter will be broadly useful to design projects aimed with inter and intra-disciplinary collaboration for sustainable cotton production.

Keywords: cotton pathogens and pests, climate change, biotic and abiotic factors, combined effect, sustainable production, IPM

1. Introduction

Cotton, *Gossypium hirsutum* is not only the most significant material yarn on worldwide basis, anyway is likewise the primary abroad exchange worker in underdeveloped nations [1]. The complete place that is known for Pakistan is 796,095 km^2, around twenty to million are developed and 8.3 million are non-developed soil. Around the globe Pakistan is among the foremost makers of cotton, wheat, sugarcane, rice, dates, mango and oranges. Our chief crops (includes cotton, wheat, sugar cane and rice) contribute 6.5% but trivial crops 2.3% of the nation's GDP. Pakistan's agriculture is based on chief crops which report for approximately 24% of the worth supplementary to generally agriculture and 4.67% of the GDP [2]. Developing countries including Pakistan have low yield production of seed cotton as compared to other technologically advanced countries like USA, Australia and China etc. Therefore, high yield of cotton is necessary on consecutive basis [3]. At present, the production of cotton yarn is approximately 25 million metric tons globally per annum basis while the current market value is around 12 billion U.S. dollars [4]. On worldwide basis, four major cotton species are cultivated to the known area for cotton production about 95%, but the key specie is *G. hirsutum* (upland cotton) that is dominant to other species as well [5]. The fast growing world population definitely demands high production of food and fiber [6]; to meet these requirements intensive agricultural practices have directed too many climatic issues importantly depletion and exhaustion of water and soil resources [7–10]. Effective agronomic practices, optimal climate and integrated pest management leads to fruitful crop production; as cotton is a perennial shrub from subtropical origin, it could even withstand harsh environmental factors e.g. drought and extreme heat [11]. Various biotic and abiotic factors are involved in reduction of cotton yield due to which cotton production significantly reduced [12]. Nowadays, breath taking climatological changes resulted in extreme weather that ultimately having direct impact on cotton production; while as plant growth reduced by facing various types of stresses in the field conditions. The dominant stresses include weather and soil factors; e.g. (temperature, humidity and rainfall) and (soil biology, altered physical and chemical properties). Generally these factors are consisted of two type's biotic and abiotic factors; biotic factors includes pathogens, weeds and pests while abiotic factors consists of temperature, humidity and drought etc. [13]. Various biotic factors such as pests and pathogens have direct effect on the cotton production and caused significant reduction in cotton crop yield estimated up to 10 to 30% [14]. While as abiotic factors such as temperature, humidity and drought etc. are even worse than biotic stresses and could cause 50% reduction [13]. GM cotton (BT cotton) was first developed in the mid-1990s and notorious for its adaptability, expressing his success story but still agronomic practices and soil system management is key spot to lessen the stressful field conditions. For sustainable productivity of cotton crop the interaction between genetics and crop agronomy is a worthy point [15].

In this chapter, our main focus is on the role of biotic and abiotic factors for sustainable cotton production. Here, we would also like to explore and summing up information about management of those factors for betterment of cotton production. We will also evaluate and identify some main reviews about the role of these factors in relation to plant response, their manifestation etc.

2. Biotic factors (biotic stress)

Biotic factors comprise chiefly insect pests and pathogens that are mainly responsible for reduction of cotton yield [14, 16, 17]; generally accounting the production losses of cotton yield due to insect pests around 84% [17] while for diseases and pathogens up to 30% [14]. The main biotic factors that leads to biotic stress for cotton field include insect pests (chewing and sucking pests) and some pathogens (viruses, bacteria and fungi) and weeds. While at the same time the abiotic factors particularly affect the occurrence of these pests and diseases [18]. Consequently, this situation is accountable for the emergence and incidence of the new pests and pathogens. Coupled with this we are viewing the main biotic factors, their key characteristics and incidence mechanisms and damage modes.

2.1 Categorization of major pests and pathogens of cotton crop

2.1.1 Insect pests

In the world over thirteen hundred and twenty six type of pests that mainly insects found as basis of infestation in cotton [19], whereas 162 different types of insect and mite pests in sub-continent [20]. The cotton crop is attacked by 96 different types of insect pests and mite species during its growth period. These insects and other pests normally observed to nourish on fiber crop in Pakistan [21]. Mainly insect pests of cotton crops consisting of two groups depending upon their mode of damage, insects with chewing mouth parts and insects with piercing sucking type of mouth parts [22]. Chewing insects generally chewed the different parts of cotton plant including cotton bollworms and leaf worms etc. e.g. pink bollworm, American bollworm, armyworm, spotted bollworm, cotton leaf worm etc. Mainly the immature stage of this group damages to cotton crop [14, 22]. **Table 1** includes the major chewing pests of cotton crop as given below.

The other group of cotton pests include pests with sucking mouth parts. In Pakistan, India, China, Japan, Trinidad, Puerto Rico, Australia and Europe the key pests of cotton crop had sucking type of mouth parts are thrips, jassid, whitefly and they cause considerable loss in cotton production [35, 36]. Insects with sucking type of mouth parts became primary pests of cotton crop from the time transgenic cotton (Bt cotton) commercialized in the world [37]. **Table 2** includes the major piercing sucking pests of cotton crop as given below.

2.1.2 Pathogens

Different fungi, bacteria and viruses are the major pathogens in cotton farming system [14]. Bacterial blight is one of the notorious disease of cotton crop caused by bacteria, *Xanthomonas citri* pv. Malvacearum; most damaging and vital cotton

Common Name	Scientific name, Order: Family	Damage	Reference
Pink bollworm	*Pectinophora gossypiella* (Saunders) (Lepidoptera: Gelechiidae)	Larvae feed on flower buds, flowers, bolls and the developing seeds within damaged bolls, resulting in the boll rotting, premature or partial boll opening also affects lint quality	[23, 24]
Old world bollworm	*Helicoverpa armigera* (Hubner) (Lepidoptera: Noctuidae)	Larvae feeds on reproductive parts of cotton, larvae also eat vegetative parts when reproductive parts are not available	[25, 26]
Corn earworm	*Helicoverpa zea* (Boddie) (Lepidoptera: Noctuidae)	Fruiting structures are damaged; feeding facilitates entry of diseases, Young shoots and leaves are damaged, in the absence of fruiting parts	[27, 28]
Spiny bollworm	*Earias insulana* (Boisdual) (Lepidoptera: Nolidae)	Larvae feed on bolls and buds of cotton, and cause. Heavy damage decreases the quantity, quality and fiber length	[29, 30]

Common Name	Scientific name, Order: Family	Damage	Reference
Spotted bollworm	*Earias vitella* (Fabricius) (Lepidoptera: Nolidae)	Larvae usually appear at the fruiting stage on cotton, damages buds, shoots,	[31, 32]
Fall armyworm	*Spodoptera frugiperda* (J. E. Smith) *Spodoptera exigua* (Hübner) *Spodoptera littoralis* (Boisd.) (Lepidoptera: Noctuidae)	Infestations occur sporadically across production regions and within fields. Neonates typically exhibit gregarious feeding behavior on the leaf near the site of oviposition, older larvae can feed within reproductive structures	[33, 34]

Table 1.
Major chewing pests of cotton.

Common Name	Scientific name, Order: Family	Damage	Reference
Lygus bug	*Lygus Hesperus* (Knight) Hemiptera; Miridae	Pierce squares, damage anthers and other tissues, bolls shrivel, turn brown, and drop from the plant	[38–40]
Green mired	*Creaontiades dilutes* (Stal) Hemiptera; Miridae	Both adults and nymphs feed on meristem tissue, loss of squares, damage growing tips, immature balls and delay maturity ultimately delay growth	[41–43]
Mirid plant bug	*Adelphocorus saturalis* (Jakovlev) Hemiptera, Miridae	Fed on newly formed tissues, shedding of squares, immature bolls and delayed maturity	[41, 43, 127,]
Stink bug	*Nezara viridula* (L.) Hemiptera, Pentatomidae	Abscise small bolls and damage lint quality, seed germination and overall yield	[44–46]
Dusky cotton bug	*Oxycarenus faetus*, (F.) Hemiptera, Lygaeidae	Suck sap from immature seeds and reduce lint quality, reduce weight and viability of seeds	[47]
Whitefly	*Bemicia tabaci* (Gennadius) Homoptera, Aleyrodidae	Sucking of sap, transmission of various viruses, honeydew associated with fungal growths	[47]

Common Name	Scientific name, Order: Family	Damage	Reference
Jassid/Leafhopper, leafhoppers	*Amrasca biguttula biguttula* (I.) *Austroasca viridigrisea* (P.) *Amrasca terraereginae* (P.) Hemiptera, Cicadellidae	Ingest cell sap from cotton leaves and inject toxic substances, leaf curling, stunting growth, distortion of plants	[48–50]
Aphid	*Aphis gossypii* Glover Hemiptera, Aphididae	Ingest cell sap from leaves and transmit viral diseases, secrete honeydew that cause sooty mold	[49]
Thrips	*Thrips tabaci* (L.) *Frankliniella occidentalis* (P.) *Frankliniella schultzai* (T.) Thysanoptera, Thripidae	Feed on terminal growth and underside of leaves, lower surface shows silvery appearance	[51, 52]
Mealy bug	*Phenococcus solenopsis* (T.) *Pseudococcus corymbatus* (M.) *Pulvinaria maxima* (G.) Saissetia nigra (N.) Hemiptera, Pseudococcidae	Suck sap from leaves, flower buds, twigs even from stem, reduce chlorophyll and water contents	[53, 54]

Source: Cotton Production, First Edition. Edited by Khawar Jabran and Bhagirath Singh Chauhan. © 2020 John Wiley & Sons Ltd. Published 2020 by John Wiley & Sons Ltd.

Table 2.
Major piercing sucking pests of cotton.

diseases caused by bacteria [55]. Bacterial blight affected the cotton crop in worst manner resulting in defoliation of the plants, stems becomes dark and swollen and even detached that ultimately leads to decrease of fiber quality, decreased coloration and death of plant [56]. Fungal diseases of cotton crop are also much more dynamic caused by various fungal pathogens, for example, *Fusarium oxysporum* is responsible to cause Fusarium wilt of cotton [57], *Verticillium dahlia* cause Verticillium wilt of cotton [58], *Colletotrichum gossypii* cause anthracnose of cotton while *Colletotrichum gossypii* var. cephalosporioides is reason for ramulosis [59], *Mycosphaerella areola* source root ramularia gray mildew [60]. The root rots of cotton crop are started by *Sclerotium rolfsii* and *Rhizoctonia solani* [61], *Alternaria macrospora* leads to leaf blight disease [62], while leaf spot of cotton caused by *Cercospora gossypina* [63] and the target spot of cotton caused by *Corynespora cassiicola* [64]. The viral diseases of cotton crop mainly includes viruses from Begomovirus family, Begomovirus acts as foundation for disease spread in cotton crop which commonly includes curl disease, leaf crumple and cotton mosaic disease [65]. While as cotton blue disease and atypical cotton blue disease introduced by Polerovirus [66, 67]. Viral diseases of cotton crop are insect transmitted and need vector for transmission in general [65].

3. Abiotic factors (abiotic stress)

Climate is changing globally nowadays because of extensive crucial human activities, while Abiotic stresses are a direct consequence of climate change. Whereas as it's foreseen to be more breathtaking increase then the average universal temperature [68]. Abiotic factors are of particular importance in regard to sustainable cotton production, because from sowing to harvest all the development stages of cotton plant depend upon abiotic factors for optimum level to grown normally. Abiotic stresses (extreme temperature, relative humidity etc.) cause distress to cotton crop and resulted in substantial decrease in yield of cotton and damage the lint quality [69, 70]. Yield loss in cotton due to the influence of abiotic factors is around 50% [13]. The major particular characteristic of abiotic tensions is their occurrence and acquaintance at the same time, simply they act same upon plant growth and plant defense behaviors that makes it more complex [71]. Keeping in view the roles of abiotic factors (abiotic stresses), here, we will analyze and summarize the effect of those in relation to sustainable cotton farming system.

3.1 Temperature impact

All crops depends upon the optimum temperature for the proper growth and development, so temperature is a chief dynamic for farming of different crops. The growth rate of cotton crop mainly dependent on the temperature, although the cotton plant is basically from tropical origin and can survive in high temperature environments but extremely hot climatic situation resulted in reduction of cotton yield [72]. Extreme temperatures either its cold or hot can ultimately affect the growth or development of cotton crop and could delay or even reduce the cotton production. Minimal changes from optimal conditions could severely effect plant growth, even 2–3°C increase from optimum temperature reduce plant growth and cause decline in yield and biomass along with increase of fiber micronaire [69]. High temperature directly or indirectly affects the growth and development of cotton and leads to drought stress because of higher evaporation from the fields. A study from Arkansas USA proposed the negative correlation between high temperature ranges and cotton production [73]. They found that the high temperature unswervingly disturbs the biology (vegetative ad reproductive stages) of cotton. This influence eventually leads to increase the production of more vegetative portion and lowering the reproductive portion of cotton plants [74]. Particularly reproductive stage of cotton crop also be strongly influenced by some other heat stress related other severe effects; in general the maturing of reproductive parts could be impacted more and cause reduction in the yield. If the temperature is higher than optimum or regular temperature, the first square formation or flowering start and first boll formation or boll opening rate decreased. Retention phase for the squares and bolls mostly decreased under high temperature [75]. Same as high temperature, low temperature also badly effect cotton growth. The development of cotton could be delayed by extreme drops of temperature under 11°C that further resulted into prolonged growth period [16]. Yadav discusses plant reactions to cold stress as well as physiological reactions and strategies for cold tolerance [76]. While at present the main focus of authors for sustainable cotton production is on high temperature under changing climatic situations.

3.2 Drought and salinity effects

Globally changing climate resulted in increase of mean temperature on per annum basis that ultimately leads to more evaporation of water from the soils and directly causing drought. Osmotic condition refers to drought and salinity stress [71]. At present, drought stress is considered among top factors that are responsible for reduction in cotton yield worldwide. Scientists foreseen that the agricultural fields strongly impacted by the drought stress will be double in the current era [77]. They also predicted that the urban, industrial and agricultural lands will face the water short comings and their will be strong competition for the water resources [78]. Cotton production strongly impacted by drought and salinity stress; drought or salinity are the main reason for reduction in cotton yield damage to the lint quality [79, 80]. Cotton plants show notable changes in their morphological and physiological characters in response so severe abiotic stresses. All the parameters including from seed germination to bolls development are impacted negatively by above stresses [69, 79, 80]. Furthermore, if the cotton crop experience those abiotic stresses for long periods of time, ultimately it faces the reduction in yield and damage to lint eminence by decreasing the boll's mass and fiber development. Thus, it faces already negative impact, also its catabolic processes like photosynthesis also got disturbed that lead to reduction in yield [69, 79, 80]. In general, plants are adaptable to drought stress or other water scarcity conditions by a variety of different internal processes, but most importantly the closing of stomatal glands is first response by numerous crops [78]. Farooq et al. suggested the comprehensive guide on the effect of drought against different crops and their physio-biochemical response [81]. Luis et al. described different biotic and abiotic stresses that could affect plant health (**Figure 1**) [82].

3.3 Combined effect of abiotic factors

The typical peculiarity of abiotic factors or abiotic stresses is that they affect the plants concurrently and causing a lot of internal and external changes simultaneously [71]. As above described cotton crop is indeterminate in nature, they have a series of systems to adapt the stressful environment. Thus, in stressful environment, for example drought or salinity problem, high temperature or soil with less minerals and nutrients, cotton plant change their behavior to face those critical environment by cracking and detaching the bolls, squares and flowers etc. [83]. Thus, under such stressful circumstances the growth of cotton roots is also limited that could be

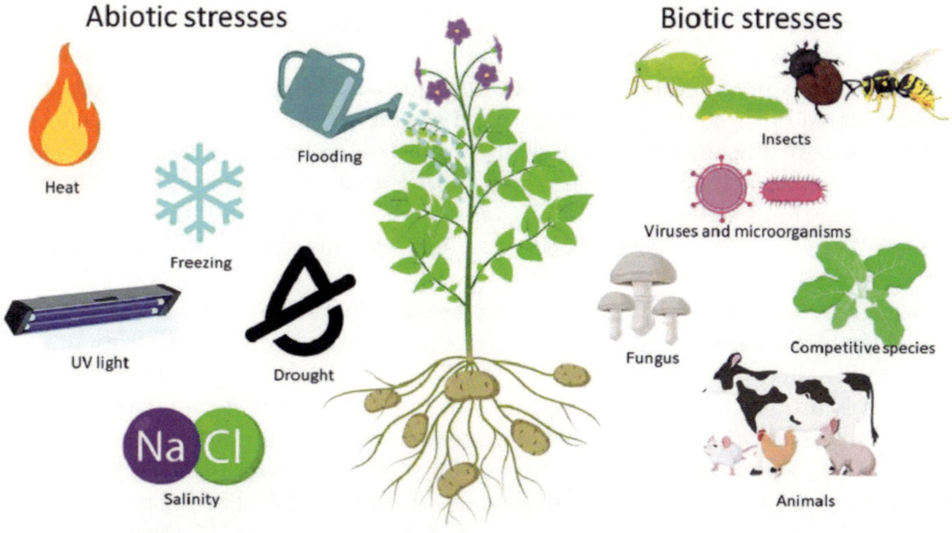

Figure 1.
Different types of biotic and abiotic stresses that can affect plants. Source: Luis et al. [82].

resulted in low yield [84]. The soils that affected by drought stress have many changes in their physiochemical properties resulting in cracking in the upper portion, superficial crusting and altered soil structure. Thus, in some cases if the coil clay content lowers to minimal level and the kind of clay micronutrients present in it sometimes leads to more than 1 m deeper cracking in soil surface. The lower level of organic matrix in soils lead to soil crusting by damaging soil hydrology [85]. The altered soil structure imparts negative effects on the interaction of soil, plant and water ultimately leading to low efficient irrigation [86]. This caused the deprived development of root network that decreases the biomass production [87].

3.4 Combined effect of biotic and abiotic factors

Various biotic and abiotic factors are involved in reduction of cotton yield due to which cotton production significantly reduced [12]. On worldwide basis, promptly increasing rate of population highly demands the food and fiber needs to be covered by the rigorous agro-farming practices [6]. But simultaneously this situation have become the big cause of global climatic change that lead to severe environment impacts particularly the water and soil resources deployment [7–10]. This state along with stark in weather measures ultimately affecting the development and growth of crops due to various kinds of stressful field condition at the same time including biotic and abiotic stresses.

The insects are highly dependent on abiotic factors and type of crop for their proper growth and development. The population growth rate is very much during suitable environmental situations and populace abundance is too high. Temperature and relative humidity plays a role in affecting the incidence of these insect pests [88]. Temperature plays a key role in controlling the development of insects and outbreaks of their population [89]. The positive relationship has been found between the population of cotton thrips and environmental factors including degree of hotness, relative humidity and rainfall [90, 91].

Rigorousness of different diseases directly associated with individual functional state of plant, the type of medium or soils, diverse changing weather situations, alike temperature, humidity or rainfall. E.g. during the wet weather condition in sandy type of soils the wilt diseases of cotton occurred more. Same as, high level of moisture and low temperature situation highly favor the growth of blight disease triggered by the *P. exigua* [92]. While on the other hand, severely cold climatic condition or chilling pressure directly influence the vulnerability of cotton crop by the attack of *A. macrospora* [93]. Likewise cold weather with low temperature range and high humidity level favor the diseases caused by *R. solani* and hot weather condition with high degree of hotness with low level of humidity leads the infection by *R. bataticola* in cotton crop [94].

High degree of hotness have direct impact on the sustainable production of cotton crop while indirect effect is increasing the rate of transpiration of water from the soil that ultimately led to drought stress [95]. Thus, the health of soil or soil condition directly influence the relative degree of drought and hotness stress the plants. While as the lower stomatal activity by plants was observed under highly fertile soils in comparison with low fertile soil [96]. Suzuki et al. described the complete survey of combined biotic and abiotic stress effect to cotton crop [97]. Pandey et al. comprehensively reviewed the Impact of Combined Abiotic and Biotic Stresses on Plant Growth and Avenues for Crop Improvement by Exploiting Physio-morphological Traits (**Figure 2**) [98].

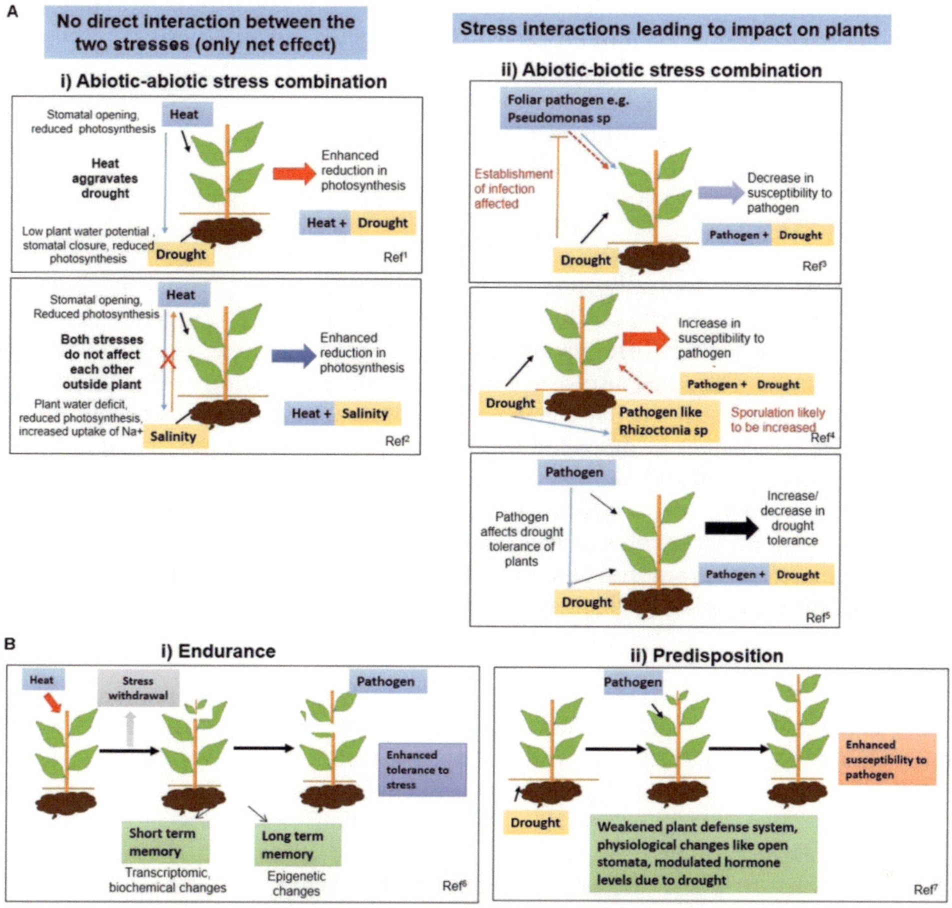

Figure 2.
Schematic representation of effect of stress combinations on plants. Source: Pandey et al. [98].

4. Management of biotic factors

In the current scenario, chemical control is the main tactic for pest and disease control, mostly farmers rely on use of synthetic insecticides, fungicides, bactericides and nematicides etc. to reduce pest infestation and disease control. On worldwide basis, to control or manage the pests and diseases conventional chemicals used. Bu however the non-judicious or inappropriate use of these synthetic pesticides have severe impacts on climate overall by damaging human and animal health, environmental degradation, resistance problem among insect pests and killing non target insects. Thus, in this regard it is necessary to design some ecofriendly and effective pest and pathogen management techniques [99] by following pre planned IPM program.

4.1 Integrated pest management

IPM simply refers to management of pests by using multiple control tactics with manipulation according to needs, that is environment friendly and effective management program for agro-ecosystems. The chemical control is the final step in

IPM. Because many insecticides have severe environmental hazards, they could have impact on beneficial insects, human health and ecosystem degradation, so in this program selective products could be used. IPM program basically have three major parts; first is the avoidance of pests; second is post sampling and monitoring; the conservation of beneficial factors and other is use of chemical and non-chemical methods to control pests in efficient way [100–102].

4.2 Pest's avoidance

For pests avoidance basically one must have comprehensive information about the behavior of particular insect pest, its ecology and biology. This tactic is broader in range and effective against different commodities according to area based pest management programs. The basic purpose of it's to keep the population of that pest below threshold level, because if the population of that particular pest is above economic threshold, the crop would be damaged by unrecoverable loss. It is important in this tactic to must study well and identify the natural enemies of that pest and also alternative hosts etc. Then management tactics could be designed according to need its set basically the foundation of effective management program. The control measures must include use of pheromones or semio-chemicals and other bio-pesticides in combination for mass trapping by disturbing reproduction cycle of that pest [103, 104].

4.3 Cultural management

For sustainable cotton production cultural control of pests is also of equal importance according to agro based region and climate. Before the sowing of genetically modified cotton, BT cotton, delayed planting in Pakistan was used to manage the pink bollworm to reduce its infestation [105]. Late season planting of cotton in Australia was avoided because of low yield and more vulnerability of cotton to be attacked by late season outbreak of sucking pest and bollworms. The destruction of pupae was followed by plowing to control *Helicoverpa armigera* [106]. Hussain et al. found high abundance of thrips on early planted cotton in Pakistan. They suggested that because the temperature have positive role in development of thrips and other sucking pests, so, by managing the agronomic factors cotton yield could be enhanced by reducing the infestation of thrips [107]. Regarding these factors pre-planned IPM practices could be designed for effective pest management.

4.4 Crop rotation

Crop rotation is also a conventional method use by many farmers on large scale cotton farming areas. But in different cotton farming systems, crop rotation have both negative and positive impacts on biotic stress factors according to the crop selected [10, 108]. To destruct the disease cycle and pest chain, choice of most appropriate crops for rotation purposes is key in integrated disease management (IDM), however, the release of disease suppressive microbes to dominate the disease are yet a challenge due to various factors [109].

4.5 Host plant resistance (HPR)

Resistance among different kinds of insect pests is an emerging challenge in effective pest management crop as cotton plants confer it due its several traits. While some varieties of cotton that have high leaf hair structure and less gossypol glands are somehow resistance against thrips, although exact mechanism is not clear. Same as some cotton cultivars with okra shaped leaves showed partial resistance against whitefly [106]. So by increasing host plant resistance could be effective as well to manage insect pests. In this regard, [14], summarized detailed biotechnological solutions for sustainable cotton production.

4.6 Sampling and monitoring of pests

The fundamental step in studying population dynamics and decision making for effective pest control strategies is sampling. For different agro farming systems sampling considered to be as primary tool. Correct sampling measures at right time not only helpful in effective pest management program but also used to check and evaluate the management options and their efficiency. In Arizona, with the effective sampling tactics and their implementation, the successful control of whitefly have been achieved [102]. In present situation as Bt cotton have been grown on worldwide basis, so by following efficient sampling procedures beneficial insects (natural enemies) could be conserved as well.

4.7 Conservation of major beneficial insects

Many beneficial insects for cotton farming system have been reported from different countries including Australia that predates and parasitize on cotton pests especially hymenopterans mainly including spiders, true bugs, lacewings, some beetles and hymenopteran parasitoids etc. [110–112]. But because of limited available methods to increase their populations for their effective use as bio control agents make them hard in this scenario. But it have been found that under unsprayed cotton fields these pests generally effective to low the pest infestations [113], but it depends on the type of pests [114, 115]. The vegetative area is extremely low in many cotton growing regions including Australia [112] that ultimately showing the no provision of refuge sites for natural enemies. That's why the beneficial insects prefers to migrate to other places for their better survival, for over wintering stage to pass and less competitive food resources [111, 112, 116]. So by planning effective agro farming systems these beneficial insects could be conserved by providing them refuge crops, provision of food baits, inoculative releases, tolerating early season pest attack, resistant varieties, use of selective and appropriate insecticides etc.

4.8 Judicious insecticidal use

Use of synthetic insecticides is the last option in IPM programs, while even the use of different botanicals and other bio pesticides are also avoided to spray in early cotton season, thus the beneficial insects could be conserved. But when the pest population exceeds to economic threshold level, then by relying and designing the appropriate time and schedule of applying sprays only with selective chemicals according to their minimal impact on non-target could help to conserve beneficial insects. Avoidance by blind sprays could also help in IPM program, because the spot treatments within field could conserve natural environment for beneficial insects.

5. Management of abiotic factors

Climate changing is a global issue now and it causing a range of abiotic stresses. The notorious characteristic of abiotic factors is that they occur simultaneously in the environment. In this regard, long term exposure to abiotic stresses resulted in reduced yield ultimately. While adaptability is a major function of living beings. To fight against abiotic stresses or to minimize the effects of different abiotic factors, many plants particularly cotton have developed some defense mechanisms to overcome the impacts of abiotic stresses. These adaptabilities by cotton plants are based on physiological or molecular mechanisms. However, still an integrated management is still needed to overcome these harsh environmental stresses.

5.1 Drought stress management

Drought is among the key limitation factor to suppress sustainable cotton production all over the world. In this regard, to overcome the effects of drought meaningfully, one must know comprehensively the soil conditions by exploration deeply [117]. The growing of commercial scaled upland cotton always show positive association with adequate water supply whenever grown in dry lands [118]. Cotton growing farmers always rely on altered row configuration to minimize the effects of drought stress as basic agronomic measure [119]. While the crop rotation tactic with cereal crops improve the soil health overall by increasing its water storage capacity and accessibility to respective crop [120]. Botanists who are working on plant physiology or molecular technologists should must keep in mind these factors when they are working to develop new innovative tools for management of drought stress. However, Farooq et al. have described the comprehensive review of drought impact on plants and their physiological response [81]. It is scientifically proofed that the highly efficient approach to overcome drought effects and increase the production rate of cotton could only be attained by the appropriate or judicious use of vital natural resources water and soil [121, 122].

5.2 Temperature stress management

Climate change is a global issue now. While this challenging the environment overall by causing abiotic stresses directly. The globally increasing temperature is mainly due to excess carbon dioxide contented in the atmosphere. Thus, the mean increase in degree of hotness globally per annum basis leads to water transpiration from the soil that ultimately resulted in drought and salty soils [71]. Temperature stress directly damages to reproductive and vegetative phase of cotton resulting in low yield [74]; although cotton is a tropical crop and have the ability to absorb severe radiation from the sun, and resulting in high temperature of the crop during high solar intensity. While in this regard, some of the plants show adaptable behavior and respond to temperature stress by developing waxy surfaces to imitate the strong radiations from the sun. But the cotton plants responded by absorption of solar radiation that intensifies osmatic (drought) pressure. Varieties with some appropriate and desired features including dense cuticle structure and profound hairs to lower down the temperature or heat stress could helpful in this regard (CICR 2016). Frequent watering could help to overcome water stress. Recently, some new innovative plant based irrigation systems have been developed by means of temperature sensing remotes through canopy [95]. While the other adverse behavior from cotton plants is the closing of stomata in the night ties that does not allow the cooling effect. In this context, plant breeders and biotechnologists may focus on developing such verities or cultivars that could fight against heat stress. As the roots are vital part to supply water to whole plant, so studying their structure and role is of equal importance and the types of soils as well. Because the sandy soils have less ability to store water for long so this could also aggravate the water and heat stress.

5.3 Role of root morphology to combat above stresses

To overcome the drought effects management tactics may be applied according to soil or plant type. Soil type and state is vital to play key role to help to develop deeper, denser and stronger plant roots to increase the use of available resources. This could ultimately helpful to significantly overcome the severe impacts of heat and drought stress. A comprehensive description on the impacts of different dry lands on roots in relation to development of crop was covered by Whitmore and Whalley [117]. Roots have different morphology among top layer of soil and sub layer of soil according to its thick structure, it could have strong influence on the behaviors of plant; how could they manage different stresses. The plants with roots who have ability to grow more bottomless or deep layers in the soil fight drought stress in a better way [123]. Some latest findings from Australia confirmed that the fine root structure of Bollgard Roundup Ready® genetically modified cultivar, was less than conventional cotton. The production of fine roots was much lower [124]. Bell also found that the cotton roots acquire phosphorus from subsoil instead of topsoil. It is proposed that under drought conditions plants roots could active in bottom and deeper soil layer while inactive at topsoil [125]. But this stressful situation ultimately depleting the P resources among cereal crop soils in Australia where the cotton is also grown [126].

5.4 Role of soil condition to combat above stresses

Different biotic and abiotic factors are directly or indirectly related to occurrence of different diseases, outbreak and survival of insect pests and could lead to complex situation in relation with different crops to manage those pests and pathogens. Thus, the soil type, structure and composition also could adversely affect these factors. In this regard, the altered soil texture severely affects the plant-soil-water relationship and cause the low water uptake by plant roots [86] that ultimately leads to poor growth and development of plant [87]. The adverse impacts of high temperature, dry and saline soils under field could be lessened or by adopting many technological measures. These measures are generally comprised of latest irrigations systems, soil health betterment by different management approaches, and management of crop residues and selection of particular crop cultivar. All the above approaches are usable but not limited to overcome the biotic and abiotic stress. Tillage have facility to alter the soil structure in favor to keep balance between hotness and soil humidity level. Thus, tillage have direct effect on transpiration of water from soil to air and infiltration through deep soil. So, some previous findings have supported that the soils with zero or minimum tillage have bettered structure to fight these stresses by helping undisturbed root development [127]. However, in some specific conditions due to use of heavy machinery on farmland resulted in hard or compacted subsoil layer needs deep tillage for better soil structure for farming. Summing up, the integrated management of biotic and abiotic stresses entails attentive use of available tactics and strongly demands development of stress tolerant varieties by keeping in mind the

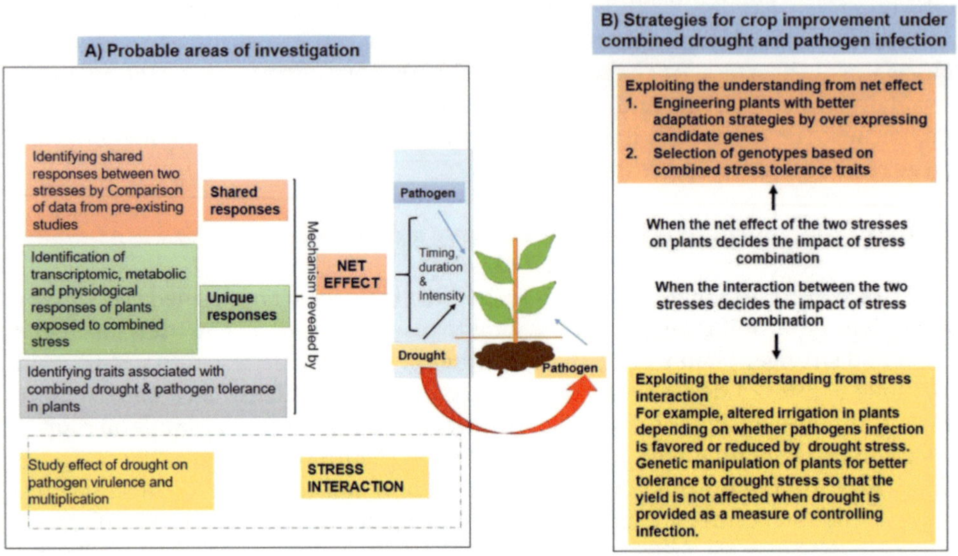

Figure 3.
Outline of strategies for improving crop performance under combined drought and pathogen stress. Source: Pandey et al. [98].

above described factors. While as Pandey et al. suggested in detail to overcome the combined effect of these stresses (**Figure 3**) [98].

6. Conclusion

Globally changing climate ultimately resulting in adverse agro-eco environments. Presently, the multi-dimensional and concurrent impacts of different biotic and abiotic factors limiting the crop growth. Raising of crop nowadays facing multiple stresses at the same time in the field. It makes ultimately hard to find tangible solutions to stand against this situation. Nevertheless, Sustainable management techno-solutions will be necessary to meet the societal requirements in this regard in the cotton-growing areas of the world. Reduction in yield of cotton crop due to multiple stresses in the field condition occurred. Different abiotic factors favoring biotic factors (insects and pathogens) to grow and survive. These factors are intermingled at some point and be reason for strong reduction in cotton yield. Co-occurrence of many stresses affects more to a plant as compared to single stress factor. Integrated pest management, integrated soil and crop management and effective agronomic practices could play a key role to combat the adverse effects of biotic and abiotic factors. Different molecular breeding methods are often used to develop cotton varieties that are resistant to abiotic stresses, whereas various transgenomic techniques mends resistance against different insect pests and pathogen. So, Plant breeders, plant physiologists, and microbiologists should must ponder over this complex situation when they forecasting their investigation approaches. Research on new innovative combinations of agro-eco systems and introduction of bio-techno solutions and persistent stewardship from farm to global farming culture is vital for sustainable cotton production.

Author details

Mehboob Hussain, Xi Gao, Deqiang Qin, Xiaoping Qin and Guoxing Wu*
State Key Laboratory for Conservation and Utilization of Bio-Resources in Yunnan, Yunnan Agricultural University, Kunming, China

*Address all correspondence to: wugx1@163.com

References

[1] Morris D. The importance of cotton to developing countries. In: Griffon M, editor. Economie des Filières en Régions Chaudes: Formation des Prix ET Échanges Agricoles. Montpellier: Acte du Xe Séminaire d'Économie et de Sociologie, 11-15 September 1989; 1990. pp. 135-146

[2] Azam A, Shafique M. Agriculture in Pakistan and its impact on economy—A review. International Journal on Advanced Science and Technology. 2017;**103**:47-60

[3] USDA. Cotton: World Markets and Trade. New York, USA: United States Department of Agriculture, Foreign Agricultural Service, Circular Series, FOP; 2010. pp. 1-10

[4] Khan MA, Wahid A, Ahmad M, Tahir MT. World cotton production and consumption: An overview. In: Ahmad S, Hasanuzzaman M, editors. Cotton Production and Uses. Singapore: Springer Nature Singapore Pte Ltd; 2020. pp. 1-7. DOI: 10.1007/978-981-15-1472-2_1

[5] Trapero C, Wilson IW, Stiller WN, Wilson LJ. Enhancing integrated pest management in GM cotton systems using host plant resistance. Frontiers in Plant Science. 2016;**7**:500

[6] FAO. World Agriculture: Towards 2015/2030. Rome, Italy: Economic and Social Development Department of FAO; 2002

[7] Bunzel K, Schäfer RB, Thrän D, Kattwinkel M. Pesticide runoff from energy crops: A threat to aquatic invertebrates? Science of the Total Environment. 2015;**537**:187-196

[8] Drinkwater LE, Snapp SS. Nutrients in agroecosystems: Rethinking the management paradigm. Advances in Agronomy. 2007;**92**:163-196

[9] Hart MR, Quin BF, Nguyen ML. Phosphorus runoff from agricultural land and direct fertilizer effects. Journal of Environmental Quality. 2004;**33**:1954-1972

[10] Nachimuthu G. Crop Rotation and Soil Health. Spotlight Magazine. Narrabri NSW, Australia: Cotton Research and Development Corporation; 2016. p. 19

[11] Arshad Awan Z, Khaliq T, Masood Akhtar M, et al. Building climate resilient cotton production system for changing climate scenarios using the DSSAT model. Sustainability. 2021;**13**:10495. DOI: 10.3390/su131910495

[12] Anonymous. 2005. U.S. Cotton Market Monthly Economic Letter, Cotton Incorporated Strategic Planning. North Carolina, USA: Cotton Incorporated; February-2005

[13] Boyer JS. Plant productivity and environment. Science. 1982;**218**:443-448

[14] Tarazi R, Jimenez JL, Vaslin MF. Biotechnological solutions for major cotton (Gossypium hirsutum) pathogens and pests. Biotechnology Research and Innovation. 2020;**3**:19-26. DOI: 10.1016/j.biori.2020.01.001

[15] Fischer RA. Chapter 2 – Farming systems of Australia: Exploiting the synergy between genetic improvement and agronomy. In: Crop Physiology. San Diego: Academic Press; 2009. pp. 22-54

[16] Bange M, Baker JT, Bauer PJ, et al. Climate Change and Cotton Production in Modern Farming Systems. Boston: CAB International; 2016. p. 61

[17] Kamburova V, Abdurakhmonov IY. Overview of the biosafety and risk assessment steps for insect-resistant biotech crops. In: Emani C, editor. The Biology of Plant-Insect Interactions. In a Compendium for the Plant Biotechnologist. USA: CRC Press; 2018. pp. 178-203. DOI: 10.1201/9781315119571

[18] URR M, Qasim M, Bukhari SA, Shaheen T. Chapter 6 – Bt crops: A sustainable approach towards biotic stress tolerance. In: Emerging Technologies and Management of Crop Stress Tolerance. San Diego: Academic Press; 2014. pp. 125-142

[19] Atwal AS. Agricultural Pests of South Asia and their Management. Ludhiana, India: Kalyani Publishers; 2002. p. 221

[20] Manjunath, TM. Bt cotton in India. In: The technology wins as the controversy wanes. 2004. Available from: www.Monsanto.co.uk/newsukshowlib.html?Wid=8478.

[21] Younas M, Yousaf M, Jilani G. Insects and Spider Mite Pests of Cotton in Pakistan. Faisalabad, Pakistan: Monog. PL-480 Dept. Entomol. Univ. Agric; 1980. p. 256

[22] Anees M, Shad SA. Insect pests of cotton and their management. In: Ahmad S, Hasanuzzaman M, editors. Cotton Production and Uses. Singapore: Springer Nature Singapore Pte Ltd; 2020. pp. 177-212. DOI: 10.1007/978-981-15-1472-2_11

[23] Lykouressis D, Perdikis D, Samartzis D, et al. Management of the pink bollworm Pectinophora gossypiella (Saunders)(Lepidoptera: Gelechiidae) by mating disruption in cotton fields. Crop Protection. 2005;**24**:177-183

[24] AKH S, Abd-El Rahman A. Relative toxicity of some modern insecticides against the pink bollworm, Pectinophora gossypiella (Saunders) and their residues effects on some natural enemies. International journal of Environmental Science and Technology. 2014;**3**:481-491

[25] King, A.B.S. Helicoverpa armigera (cotton bollworm). 2106. Available from: http://wwwcabiorg/isc/datasheet/26757 (CABI ed)

[26] Kranthi S, Kranthi K, Wanjari R. Influence of semilooper damage on cotton host-plant resistance to Helicoverpa armigera (hub). Plant Science. 2003;**164**:157-163

[27] Jackson R, Bradley J Jr, Van Duyn J. Field performance of transgenic cottons expressing one or two bacillus thuringiensis endotoxins against bollworm, Helicoverpa zea (Boddie). Journal of Cotton Science. 2003;7:57-64

[28] King, A.B.S. Helicoverpa zea (American cotton bollworm), (CABI ed), 2016. Available from: http://www.cabi.org/isc/datasheet/26776

[29] Mirmoayedi A, Maniee M, Yaghutipoor A. Control of cotton spiny bollworm, Earias insulana Boisduval, using three bio-insecticides, Bt, Spinosad and neem-Azal. Journal of Entomology. 2010;7:89-94

[30] Hajatmand F, Abbasipour H, Amin G, et al. Evaluation of infestation percentage of cotton fields to the spiny bollworm, Earias insulana Boisduval (Lep.: Noctuidae), and its relationship with pheromone traps. Archives of Phytopathology and Plant Protection. 2014;**47**:1523-1529

[31] Ahmad M, Iqbal Arif M. Resistance of Pakistani field populations of spotted bollworm Earias vittella (Lepidoptera: Noctuidae) to pyrethroid, organophosphorus and new chemical

insecticides. Pest Management Science. 2009;**65**:433-439

[32] Hasan W, Ansari M. Evaluation of some insecticides against spotted bollworm, Earias vittella (fab.) on different okra cultivars. Trends in Biosciences. 2010;**3**:41-44

[33] Reed J, Smith D. Droplet size and spray volume affects on insecticide deposit and mortality of Heliothine (Lepidoptera: Noctuidae) larvae in cotton. Journal of Economic Entomology. 2001;**94**:640-647

[34] Hardke JT, Jackson RE, Leonard BR, Temple JH. Fall armyworm (Lepidoptera: Noctuidae) development, survivorship, and damage on cotton plants expressing insecticidal plant-incorporated protectants. Journal of Economic Entomology. 2015;**108**:1086-1093

[35] Amjad A, Aheer GM. Varietal resistance against sucking insect pests of cotton under Bahawalpur ecological conditions. Journal of Agricultural Research. 2007;**45**:1-5

[36] Aslam M, Razaq MM, Saeed NA, Ahmad F. Comparative resistance of different cotton varieties against bollworm complex. International Journal of Agriculture and Biology. 2004;**6**:39-41

[37] Holt J, Mushobizi W, Day RK, Knight JD, Kimani M, Njuki J, et al. A simple Bayesian network to interpret the accuracy of armyworms outbreak forecasts. The Annals of Applied Biology. 2006;**148**:141-146

[38] Cervantes FA, Backus EA, Godfrey L, et al. Characterization of an EPG waveform library for adult Lygus lineolaris and Lygus hesperus (Hemiptera: Miridae) feeding on cotton squares. Annals of the Entomological Society of America. 2016;**109**:684-697

[39] Meisner MH, Zaviezo T, Rosenheim JA. Landscape crop composition effects on cotton yield, Lygus hesperus densities and pesticide use. Pest Management Science. 2017;**73**:232-239

[40] Zhang L-J, Cai W-Z, Luo J-Y, et al. Phylogeographic patterns of Lygus pratensis (Hemiptera: Miridae): Evidencee for weak genetic structure and recent expansion in northwest N China. PLoS One. 2017;**12**:e0174712

[41] Anonymous. Mirids, Green, Brown and Crop. Queensland: Department of Agriculture and Fisheries; 2010

[42] Hill L. Migration of green mirid, Creontiades dilutus (Stål) and residence of potato bug, Closterotomus norwegicus (Gmelin) in Tasmania (Hemiptera: Miridae: Mirinae: Mirini). Crop Protection. 2017;**96**:211-220

[43] Armes NJ, Jadhav DR, De Souza KR. A survey of insecticide resistance in Helicoverpa armigera in the Indian subcontinent. Bulletin of Entomological Research. 1996;**86**:499-514

[44] Cui H-H, Gu S-H, Zhu X-Q, et al. Odorant-binding and chemosensory proteins identified in the antennal transcriptome of Adelphocoris suturalis Jakovlev. Comparative Biochemistry and Physiology Part D: Genomics and Proteomics. 2016;**24**:139-145

[45] Willrich M, Leonard B, Gable R, Lamotte L. Boll injury and yield losses in cotton associated with brown stink bug (Heteroptera: Pentatomidae) during flowering. Journal of Economic Entomology. 2004;**97**:1928-1934

[46] Ahmed R, Nadeem I, Yousaf MJ, et al. Impact of dusky cotton bug (Oxycarenus laetus Kirby) on seed germination, lint color and seed weight

in cotton crop. Journal of Entomology and Zoology Studies. 2015;**3**:335-338

[47] Li J, Zhu L, Hull JJ, et al. Transcriptome analysis reveals a comprehensive insect resistance response mechanism in cotton to infestation by the phloem feeding insect Bemisia tabaci (whitefly). Plant Biotechnology Journal. 2016;**14**:1956-1975

[48] Amin M, Afrin R, Alam M, et al. Effect of leaf trichomes and meteorological parameters on population dynamics of aphid and jassid in cotton. Bangladesh Journal of Agricultural Research. 2017;**42**:13-25

[49] Bi JL, Ballmer GR, Hendrix DL, et al. Effect of nitrogen fertilizer on cotton whitefly Bemisia argentifolii populations and honeydew production. Entomologia Experimentalis et Applicata. 2001;**99**:25-36

[50] Lei T, Khan M, Wilson L. Boll damage by sucking pests: An emerging threat, but what do we know about it. In: Swanepoel A, editor. Cotton Production for the New Millenium. Cape Town: Proceedings 3rd World Cotton Conference (Cape Town, 9-13 March, 2003) CD; 2003. pp. 1338-1344

[51] Attique M, Ahmad Z. Investigation of Thrips tabaci Lind. as a cotton pest and the development of strategies for its control in Punjab. Crop Protection. 1990;**9**:469-473

[52] Kakkar G, Kumar V, Seal DR, et al. Predation by Neoseiulus cucumeris and Amblyseius swirskii on and Frankliniella schultzei on cucumber. Biological Control. 2016;**92**:85-91

[53] Prabhakar M, Prasad YG, Vennila S, et al. Hyperspectral indices for assessing damage by the Solenopsis mealybug (Hemiptera: Pseudococcidae) in cotton. Computers and Electronics in Agriculture. 2013;**97**:61-70

[54] Rahmathulla V, Sathyanarayana K, Angadi B. Influence of abiotic factors on population dynamics of major insect pests of mulberry. Pakistan Journal of Biological Sciences. 2015;**18**:215-223

[55] Jalloul A, Sayegh M, Champion A, Nicole M. Bacterial blight of cotton. Phytopathologia Mediterranea. 2015;**54**(2):241-252. DOI: 10.14601/Phytopathol

[56] Chohan S, Perveen R, Abid M, Tahir MN, Saji M. Cotton diseases. Research. 2021;**4**:25. DOI: 10.1186/s42397-021-00100-9

[57] Cox KL, Babilonia K, Wheeler T, He P, Shan L. Return of old foes—Recurrence of bacterial blight and Fusarium wilt of cotton. Current Opinion in Plant Biology. 2019;**50**:95-103. DOI: 10.1016/J.PBI.2019.03.012

[58] Shaban M, Miao Y, Ullah A, Khan AQ, Menghwar H, Khan AH, et al. Physiological and molecular mechanism of defense in cotton against Verticillium dahliae. Plant Physiology and Biochemistry. 2018;**125**:193-204. DOI: 10.1016/J.PLAPHY.2018.02.011

[59] Salustiano ME, Rondon MN, Abreu LM, da Costa S, da Machado JC, Pfenning LH. The etiological agent of cotton ramulosis represents a single phylogenetic lineage within the Colletotrichum gloeosporioides species complex. Tropical Plant Pathology. 2014;**39**(5):357-367. DOI: 10.1590/S1982-56762014000500002

[60] Shete PP, Kasal YG, Perane RR. Screening of the cotton genotypes against ramularia areolaatk. Under field condition. Plant Archives. 2018;**18**(1):734-736

[61] Mehta Y, Marangoni MS, Bocatti RC, Rodrigues HP, Cunha TS, Galbieri R. Systemic acquired resistance of cotton, soybean and common bean to rhizoctonia solani and sclerotium rolfsii induced by shale water seed treatment. American Journal of Plant Sciences. 2015;**6**(9):1493-1500. DOI: 10.4236/ajps.2015.69148

[62] Cia E, Fuzatto MG, Kondo JI, Carvalho LH, Ito MF, Dias FLF, et al. Response of cotton genotypes to the incidence of Alternaria leaf spot. Summa Phytopathologica. 2016;**42**(4):357-359. DOI: 10.1590/0100-5405/2119

[63] Rothrock CS, Woodward JE, Kemerait RC. Diseases. In: Cotton. 2nd ed. Madison, WI: Agron. Monogr. 57. ASA, CSSA, and SSSA; 2015. pp. 465-508. DOI: 10.2134/agronmonogr57.2014.0071

[64] Galbieri R, Araújo DCEB, Kobayasti L, Girotto L, Matos JN, Marangoni MS, et al. Corynespora leaf blight of cotton in Brazil and its management. American Journal of Plant Sciences. 2014;**5**(26):3805-3811. DOI: 10.4236/ajps.2014.526398

[65] Naqvi RZ, Zaidi SS, Shahid Mukhtar M, Amin I, Mishra B, Strickler S, et al. Transcriptomic analysis of cultivated cotton Gossypium hirsutum provides insights into host responses upon whitefly-mediated transmission of cotton leaf curl disease. PLoS One. 2019;**14**(2):1-21. DOI: 10.1371/journal.pone.0210011

[66] Correa RL, Silva TF, Simões-Araújo JL, Barroso PA, Vidal MS, Vaslin MFK. Molecular characterization of a virus from the family Luteoviridae associated with cotton blue disease. Archives of Virology. 2005;**150**(7):1357-1367. DOI: 10.1007/s00705-004-0475-8

[67] Silva AKF, Romanel E, da Silva FT, Castilhos Y, Schrago CG, Galbieri R, et al. Complete genome sequences of two new virus isolates associated with cotton blue disease resistance breaking in Brazil. Archives of Virology. 2015;**160**(5):1371-1374. DOI: 10.1007/s00705-015-2380-8

[68] Luo Q, Bange M, Braunack M, Johnston D. Effectiveness of agronomic practices in dealing with climate change impacts in the Australian cotton industry – A simulation study. Agricultural Systems. 2016;**147**:1-9

[69] Majeed S, Rana IA, Mubarik MS, et al. Heat stress in cotton: A review on predicted and unpredicted growth-yield anomalies and mitigating breeding strategies. Agronomy. 2021;**11**:1825. DOI: 10.3390/agronomy11091825

[70] Noreen S, Ahmad S, Fatima Z, et al. Abiotic stresses mediated changes in morphophysiology of cotton plant. In: Ahmad S, Hasanuzzaman M, editors. Cotton Production and Uses. Singapore: Springer Nature Singapore Pte Ltd; 2020. pp. 341-366. DOI: 10.1007/978-981-15-1472-2_18

[71] Rani S, Kumar P, Suneja P. Biotechnological interventions for inducing abiotic stress tolerance in crops. Plant Gene. 2021;**27**:100315. DOI: 10.1016/j.plgene.2021.100315

[72] Ullah K, Khan N, Usman Z, Ullah R, Saleem FY, Shah SAI, et al. Impact of temperature on yield and related traits in cotton genotypes. Journal of Integrative Agriculture. 2016;**15**:678-683

[73] Oosterhuis DM. Yield response to environmental extremes in cotton. In: Oosterhuis DM, editor. Proceedings of the 1999 Cotton Research Meeting and Summaries of Cotton Research in Progress. Vol. 193. Fayetteville, USA:

Arkansas Agricultural Experiment Station special report; 1999. pp. 30-38

[74] Reddy VR, Reddy KR, Baker DN. Temperature effect on growth and development of cotton during the fruiting period. Agronomy Journal. 1991;**83**:211-217

[75] Reddy KR, Davidonis GH, Johnson AS, Vinyard BT. Temperature regime and carbon dioxide enrichment alter cotton boll development and fiber properties. Contribution from the Dep. of Plant and Soil Sciences, Mississippi State Univ., and the USDA-ARS Southern Regional Res. Ctr., New Orleans, LA. Mississippi Agric. and Forestry Exp. Stn. Paper no. J 9391. Agronomy Journal. 1999;**91**:851-858

[76] Yadav SK. Cold stress tolerance mechanisms in plants. A review. Agron Sustain Dev. 2010;**30**:515-527

[77] Deeba F, Pandey AK, Ranjan S, Mishra A, Singh R, Sharma YK, et al. Physiological and proteomic responses of cotton (Gossypium herbaceum L.) to drought stress. Plant Physiology and Biochemistry. 2012;**53**:6-18

[78] Tombesi S, Nardini A, Frioni T, Soccolini M, Zadra C, Farinelli D, et al. Stomatal closure is induced by hydraulic signals and maintained by ABA in drought-stressed grapevine. Science Reporter. 2015;**5**:12449

[79] Abdelraheem A, Esmaeili N, O'Connell M, Zhang J. Progress and perspective on drought and salt stress tolerance in cotton. Industrial Crops and Products. 2019;**130**:118-129. DOI: 10.1016/j.indcrop.2018.12.070

[80] Sharif I, Aleem S, Farooq J, et al. Salinity stress in cotton: Effects, mechanism of tolerance and its management strategies. Physiology and Molecular Biology of Plants. 2019;**25**:807-820. DOI: 10.1007/s12298-019-00676-2

[81] Farooq M, Wahid A, Kobayashi N. Plant drought stress: Effects, mechanisms and management. Agronomy for Sustainable Development. 2009;**29**:185-121

[82] Paramo LA, Feregrino-Pérez AA, Guevara R. Sandra Mendoza and Karen Esquivel (2020) nanoparticles in agroindustry: Applications, toxicity, challenges, and trends. Nanomaterials. 2020;**10**:1654. DOI: 10.3390/nano10091654

[83] Williams S, Bange MP. Chapter, 1, the cotton plant. In: Australian Cotton Production Manual 2015. Narrabri, Australia: CRDC and CottonInfo; 2015. pp. 8-10

[84] Karamanos AJ, Bilalis D, Sidiras N. Effects of reduced tillage and fertilization practices on soil characteristics, plant water status, growth and yield of upland cotton. Journal of Agronomy and Crop Science. 2004;**190**:262-276

[85] Overstreet LF, DeJong-Huges J. The importance of soil organic matter in cropping systems of the Northern Great Plains. University of Minnesota-Extension 2016; Web material. 2016. Available from: http://www.extension.umn.edu/agriculture/tillage/importance-of-soil-organic-matter/

[86] Loch RJ, Grant CG, McKenzie DC, Raine SR. Improving Plants' Water Use Efficiency and Potential Impacts from Soil Structure Change – Research Investment Opportunities, Final report to the National Program for Sustainable Irrigation. CRCIF report number 3.14/1. Toowoomba: Cooperative Research Centre for Irrigation Futures; 2005

[87] Wrona AF, Oosterhuis DM, McMichael B. Getting to the root of

your crop's health. Cotton Physiol Today. 1999;**10**:1-8

[88] Aheer GM, Ahmad KJ, Ali A. Role of weather in fluctuating aphid density in wheat crop. Journal of Agricultural Research. 1994;**32**:295-301

[89] Weisser W, Volkl WW, Hassell MP. The importance of adverse weather conditions for behaviour and population ecology of an aphid parasitoid. The Journal of Animal Ecology. 1997;**66**:386-400

[90] Li JZ, Zang HJ, Zang ZM. Observation on the time and spatial dynamics of the diurnal cycle of tobacco thrips. China Journal of Entomology. 1992;**18**:26-30

[91] Shah SIH. Effect of Abiotic Factors on the Incidence and Development of Sucking Insect Pests Complex on Different Advanced Genotypes of Cotton under Unsprayed Conditions [Thesis]. Faisalabad, Pakistan: Dept. Entomol. Univ. Agric; 2003

[92] Koenning SR, Abdel Alim FF, Grand LF, Phipps PM. Stem Canker on Cotton Caused by Phoma exigua in North Carolina and Virginia. Plant Disease. Nov 2000;**84**(11):1251. DOI: 10.1094/PDIS.2000.84.11.1251B. PMID: 30832181

[93] Zhao J, Li S, Jiang T, Liu Z, Zhang W, Jian G, et al. Chilling stress—The key predisposing factor for causing Alternaria alternata infection and leading to cotton (Gossypium hirsutum L.) leaf senescence. PLoS One. 2012;7:e36126

[94] Monga D, Raj S. Root Rot Disease of Cotton and its Management. CICR Technical Bulletin no: 3. Nagpur: Central Institute for Cotton Research; 2016

[95] White S, Raine SR. A Grower Guide to Plant Based Sensing for Irrigation Scheduling. Toowoomba: National Centre for Engineering in Agriculture Publication, 1001574/6: USQ; 2008

[96] Irmak S. Impacts of Extreme Heat Stress and Increased Soil Temperature on Plant Growth and Development. Lincoln: University of Nebraska; 2016

[97] Suzuki N, Rivero RM, Shulaev V, Blumwald E, Mittler R. Abiotic and biotic stress combinations. The New Phytologist. 2014;**203**:32-43

[98] Pandey P, Irulappan V, Bagavathiannan MV, Senthil-Kumar M. Impact of combined abiotic and biotic stresses on plant growth and avenues for crop improvement by exploiting physio-morphological traits. Frontiers in Plant Science. 2017;**8**:537. DOI: 10.3389/fpls.2017.0053

[99] Alavanja MCR. Introduction: Pesticides use and exposure extensive worldwide. Reviews on Environmental Health. 2009;**24**(4):303-309. DOI: 10.1515/REVEH.2009.24.4.303

[100] Mensah RK. Development of an integrated pest management programme for cotton. Part 1: Establishing and utilising natural enemies. International Journal of Pest Management. 2002a;**48**:87-94

[101] Mensah RK. Development of an integrated pest management programme for cotton. Part 2: Integrationi of a Lucerne/cotton interplant system, food supplement sprays with biological and synthetic insecticides. International Journal of Pest Management. 2002b;**48**:95-105

[102] Naranjo SE. Impacts of Bt transgenic cotton on integrated pest management. Journal of Agricultural and Food Chemistry. 2011;**59**:5842-5851

[103] Mensah RK, Moore C. Exploitation of semiochemicals for the management

of pest and beneficial insects with special emphasis on cotton cropping systems in Australia: A review. Journal of Biological Control. 2011;**25**(4):253-269

[104] Mensah RK, Gregg PC, Del Socorro AP, et al. Integrated pest management in cotton: Exploiting behaviour modifying compounds for managing cotton pests. Crops & Pasture Science. 2013;**64**:763-773

[105] Usman M, Razaq M, Shah F, et al. Impact of planting dates and insecticides treatments on pink bollworm damage to cotton bolls. In: Paper Presented in National Conference, Challenges and Opportunities to Boost the Agriculture in Changing Climate. Multan, Pakistan: Bahdur Sub-Campus, Layyah of Bahauddin Zakariya University; 2017 (29 March 2017)

[106] Wilson L, Downes S, Khan M, et al. IPM in the transgenic era: A review of the challenges from emerging pests in Australian cotton systems. Crop and Pasture Science. 2013;**64**:737-749

[107] Hussain M, Ahmad M, Iqbal M, et al. Seasonal outbreak of Thrips tabaci on different cotton cultivars in relation to abiotic factors and different sowing times. International Journal of Agriculture and Biological Sciences. Jul and Aug 2020. p. 66-75

[108] Kirby K, Smith L. Tackling verticilium. Spotlight magazine. Cotton Research and Development Corporation. Winter. 2016;**2016**:16-18

[109] Pereg L, McMillan M. Scoping the potential uses of beneficial microorganisms for increasing productivity in cotton cropping systems. Soil Biology and Biochemistry. 2015;**80**:349-358

[110] Room PM. Seasonal occurrence of insects other than Helicoverpa spp. feeding on cotton in the Namoi valley of New South Wales. Australian Journal of Entomological Society. 1979;**16**:165-174

[111] Mensah RK, Khan M. Use of Medicago sativa (L) interplantings/ trap crops in the management of the green mirid, Creontiades dilutus (Stal) in commercial cotton crops in Australia. International Journal of Pest Management. 1997;**43**:197-202

[112] Mensah RK. Habitat diversity: Implications for the conservation and use of predatory insects of Helicoverpa spp. in cotton systems. International Journal of Pest Management. 1999;**45**:91-100

[113] Clark MS, Luna JM, Stone ND, Youngman RR. Generalist predator consumption of army worm and effect of predator removal on damage in no-till corn. Environmental Entomology. 1994;**23**:617-622

[114] Murdoch MM, Cheeson J, Cheeson PL. Biological control in theory and practice. American Naturalist. 1985;**125**:344-366

[115] Wratten SD. The effectiveness of native natural enemies. In: Burn AJ, Croaker TH, Jepson PC, editors. Integrated Pest Management. London: Academic Press; 1987. pp. 89-111

[116] Landis DA, Wratten SD, Gurr GM. Habitat management to conserve natural enemies of arthropod pests in agriculture. Annual Review of Entomology. 2000;**45**:175-201

[117] Whitmore AP, Whalley WR. Physical effects of soil drying on roots and crop growth. Journal of Experimental Botany. 2009;**60**:2845-2857

[118] Constable GA, Hearn AB. Irrigation for crops in a sub-humid environment. Irrigation Science. 1981;**3**:17-28

[119] Payero J, Robinson G, Harris G, Singh D. Water extraction of solid and skip-row cotton. In: 16th Australian Agronomy Conference 14-18th October 2012. NSW: University of New England in Armidale; 2012

[120] Hulugalle NR, Scott F. A review of the changes in soil quality and profitability accomplished by sowing rotation crops after cotton in Australian vertosols from 1970 to 2006. Soil Research. 2008;**46**:173-190

[121] Gardner WR, Gardner HR. Plant production and management under drought conditions. Principles of water management under drought conditions. Agricultural Water Management. 1983;**7**:143-155

[122] Ostle NJ, Levy PE, Evans CD, Smith P. UK land use and soil carbon sequestration. Land Use Policy. 2009;**26**(Suppl 1):S274-S283

[123] Ho MD, Rosas JC, Brown KM, Lynch JP. Root architectural tradeoffs for water and phosphorus acquisition. Functional Plant Biology. 2005;**32**:737-748

[124] Hulugalle NR, Broughton KJ, Tan DKY. Fine root production and mortality in irrigated cotton, maize and sorghum sown in vertisols of northern New South Wales, Australia. Soil and Tillage Research. 2015;**146**(Part B): 313-322

[125] Bell M. Developing Soil Testing and Fertiliser Response Guidelines to Manage P, K and S Fertility for Irrigated and Dryland Cotton Cropping Systems, Final Report to Cotton Research and Development Corporation. Gatton: Queensland Alliance for Agriculture and Food Innovation, The University of Queensland; 2015

[126] Bell M, Lester D, Smith L, Want P. Increasing complexity in nutrient management on clay soils in the northern grain belt – Nutrient stratification and multiple nutrient limitations. In: 16th Australian Agronomy Conference, 14-18th October 2012. NSW: University of New England in Armidale; 2012

[127] Dighton J, Jones HE, Robinson CH, Beckett J. The role of abiotic factors, cultivation practices and soil fauna in the dispersal of genetically modified microorganisms in soils. Applied Soil Ecology. 1997;**5**:109-131

Chapter 5

Integrated Pest Management in Cotton

Mirza Abdul Qayyum, Shafqat Saeed, Waqas Wakil, Muhammad Yasin, Ahmad Nawaz, Abid Ali, Habib Ali, Unsar Naeem-Ullah, Umer Sharif, Hasan Taha, Aiman Khalid, Iqra Maryum, Zeeshan Shahid and Wania Hasan Zubairi

Abstract

Cotton, one of the most important crops in the world, is threatened by a wide range of pests that lower both yield and quality. Integrated Pest Management (IPM) has come to be a comprehensive and eco-friendly method to tackle these issues while decreasing the need for synthetic pesticides. The first section of this chapter gives a general review of cotton as an essential agricultural commodity and its significance to the textile industry. After that, it explores the origins and evolution of IPM, providing an in-depth understanding of its concepts and the necessary elements needed to successfully apply it to cotton cultivation. Biological control, cultural practices, chemical control, host plant resistance, and other pest management strategies covered by Integrated Pest Management (IPM) are highlighted, along with how best to integrate them for maximum efficacy. Use of technology (Artificial intelligence) in early identification and detection of pest is necessary because it can help in timely decision making for pest management. Analyzing the ecological effects of IPM also gives information on how to reduce the amount of chemicals in cotton fields and preserve beneficial fauna. It is an important tool for farmers, academics, and policymakers who want to increase cotton profitability and long-term viability while reducing its negative effects on the environment and encouraging a stronger agricultural ecosystem.

Keywords: sustainable cotton production, integrated pest management, cotton insect pests, sucking pests of cotton, chewing pests of cotton

1. Introduction

1.1 Importance of sustainable cotton production

Although the manufacturing of synthetic fibers has risen, cotton continues to be the most significant fiber crop due to its global importance, garnering it the nickname "Queen of the Fiber Plants." Cotton production is a major industry in more than 80 nations, accounting for 2.5% of all cultivated land worldwide. Cotton also has commercial worth because it is essential to its producers' ability to earn foreign currency [1, 2]. Because of improvements in farming practices and the development of the industry for manufacturing products made of cotton, a significant portion of the nation's workforce—roughly 40% of all workers—have jobs thanks to the cotton industry. Another source of edible oil that contributes about 60% of the total amount consumed is cotton seed [3]. Although cotton is a natural fiber, modern cotton products do not merit the name "natural fiber" due to the increased consumption of synthetic pesticides and fertilizers (applied to nearly 99% of cotton crops) to increase crop yields and improve industrial production.

More than 230 different insect species have been recorded attacking cotton crops worldwide [4]. Since cotton is the most widely grown crop and is also more likely to be attacked by insect pests, it is discovered that 16% of all pesticides used worldwide are used on a single cotton crop. Agrochemicals make up about 50% of the overall cost of seed cotton production worldwide [5]. The ecosystem and environment are being negatively impacted by the widespread use of hazardous pesticides and chemicals. As more and more pesticide-resistant species appear and change insect patterns, the widespread use of chemicals disturbs the biological equilibrium. All of this contributed to an increase in production costs. Due to the fact that these chemicals are released into the air and water, they also reduce biodiversity by causing pollution [6]. The answer to all new issues is organic cotton. In other words, no inorganic fertilizers, herbicides, insecticides, or fungicides are used when growing cotton; it is instead grown naturally. It has certification from a recognized certifying body.

The use of fungicides and pesticides is decreased in organic production systems to a level that supports a variety of life forms, preserving biological diversity in agriculture. This aids in both replenishing and preserving soil fertility. Other names for organic cotton include "natural cotton," "clean cotton," "green cotton," or "environmentally friendly cotton". In contrast to traditional agricultural techniques, which place a big emphasis on using chemicals, organic cotton production systems use biological materials. Chemical fertilizers are prohibited for the growth of organic cotton and must be substituted with organic manure, such as gypsum, cotton seed meal, fish meal, cake, leather meal, farmyard and green manure, composite, etc. Additionally, botanical herbicides and insecticides like ipomea, neem cake, etc., are frequently utilized. Conventional cotton farming uses pesticides, insecticides, and herbicides that have been chemically synthesized.

When picking organic cotton, the use of chemical defoliants is also avoided [7]. Organic cotton fabric is substantially more expensive because its output is less than that of conventional cotton. Because no hazardous chemicals are allowed in the production of organic cotton, not only do farmers benefit from a healthy environment, but environmental contamination is also avoided. Considering that chemicals do not alter the makeup of soil, they continue to be fertile. The use of simple, nontoxic dyes rather than dangerous chemicals like chlorine, toxic finishes, bleach, etc., during organic treatment processes reduces the need to treat water that has been contaminated with chemicals during cotton production [8].

2. Need for effective crop management and processing practices

Alternating the soil and water, as well as increased chemical use, such as that of pesticides and fertilizers, is having an impact on life as we realize it. In children, methemoglobinemia (blue baby syndrome) is a disease caused by nitrates, which originate from nitrogen fertilizers. In California, the amount of volatile organic compound (VOC) emissions has risen to 11 million pounds, and disorders linked to pesticides are now the third most common. Recently, 13 insecticides that kill birds have been registered for use on cotton. The most hazardous result of utilizing these herbicides is that fish and other wildlife cannot reproduce [9].

Environmental protection, financial success, and social and economic equality are the three objectives of sustainable agriculture. In this context, it is important to preserve and improve land and natural resources while also taking into account consumer safety, health, and work conditions for farmers, as well as rural residents' wants and expectations. Sustainable crop production adheres to methods that respect human rights, the environment, and local and rural populations while also raising crops in an ethical and ecological manner. Sustainable crop management is centered on agricultural practices that are maintained over an extended period of time with little or no impact on the environment.

2.1 Soil management

Organic farming demands healthy soil. The soil serves as both a medium for plant growth and a living system that maintains the soil's health and productivity. Compost, effective nutrient recycling, cover crops, and regular crop rotation can all substitute for synthetic fertilizers [10].

2.2 Weeds management

Hoes and other mechanical tools, crop rotation, effective irrigation water usage, intercropping, the use of mulches, modifying plant densities, and adjusting panting dates can all be used to restrict the growth of weeds [11].

2.3 Insect pest management

Farmers can decrease the probability of insects and harm to crops caused by birds or mammals, a variety of living things. Insect predators and crop rotate your crops, use biological pesticides like neem oil, and Pests can be managed by intercropping by the cultivators [12]. There are seven essential conditions that must be maintained in order to effectively combat pests. In that sequence, the following are listed: (i) diagnosis, (ii) combat plan, (iii) timing of application, (iv) choice of pesticide, (v) choice of instrument, (vi) application, and (vii) supervision [13]. When these are applied in the management's order throughout a plant protection process, success is quite likely to occur.

2.4 Harvesting

Particularly in developing nations, organic cotton is typically harvested by hand without the use of machines, chemicals, or defoliants, which minimizes waste.

3. Issues in sustainable cotton production

3.1 Runoff from fields

When agricultural land's surface is fully saturated and unable to absorb additional water, it is said to have an excess of water. Runoff is the phrase used to describe this water's flow to nearby bodies of water. Runoff, which not only transports extra water but also contains pesticide, fertilizer, and salt residues that were applied to the cotton crop, is caused by a variety of factors, including improper water management and inexperience with irrigation [14].

3.2 Drainage and leaching

Salinity in irrigation or soil is a significant issue for cotton farming. It affects crops' access to soil water and is comparable to water scarcity. Cotton can keep out sodium ions and is salinity resistant. As a result, there is a chance that issues will not be recognized until they have accumulated.

4. Crop management practices

4.1 Selection of suitable cotton varieties

The choice of seeds is made in accordance with the soil and climate conditions, farming methods, and estimated yields [15]. For mechanical drilling, 20–30 kg/ha of seed is needed, whereas 14–16 kg/ha is needed for ridge sowing [16]. The chosen seed should be able to withstand extremes in temperature as well as very low ones [17]. It should also be resistant to diseases like CLCuV and able to survive droughts [18]. The cotton varieties (CYTO 177, FH 326, BS 15, Sahara 120, Weal AG Shahkar, NIAB 878, Tarzan 3, NIABG 4, CYTO 179, CIM 600, and BPC 11) that Punjab Seed Council just issued have resistance for a variety of diseases and insect pests.

4.2 Seed treatment

The cotton seeds frequently have low germination rates. To increase germination, cotton seed is cleaned of lint [19]. Acid delinting and mechanical removal are the two processes that are most commonly employed to remove lint from cotton seed. For seed delinting, sulfuric acid and hydrochloric acid are frequently employed [20]. The heat created by the acid and moisture interaction causes the short fibers on cotton seeds to deteriorate [21]. Water is used to wash the residue. The wet acid technique, which uses soda ash or ammonia to neutralize the seed lime, is so named [22]. Fungicides are used in the treatment of delinted seed.

4.3 Sowing time

The season for planting cotton begins in April and lasts through the end of May [23]. The quality of cotton fiber is impacted by late planting. Cotton planting may begin in May in warm places (mainly in the south). To investigate a cultivar's production potential in each area, the date of planting is crucial [24]. Warm soil temperatures are necessary for cotton seed germination because they promote strong seedling growth [25]. Low soil temperature, which causes poor plant development, makes crops seeded very early, particularly in the month of February, appear to have a weak crop stand [26]. Similarly, cotton seedlings planted too late are shaded by weeds [27].

Weed infestation, CLCuV infection, and heat stress all limit crop development [28]. Any inadequate management of cotton that was planted in the late season reduces production. Accordingly, the planting date is a crucial determinant in the potential production and healthy plant growth [29].

5. Land preparation and fertility management

5.1 Land preparation

To guarantee uniform planting depth, cotton needs a surface of soil with a very fine tilth and a smooth texture [30]. To achieve a sufficient plant stand, several field operations are required. A deep soil is necessary for healthy root growth since cotton is a crop with deep roots [31]. Land needs to be clear of clods and stubble. Cotton should be deep-plowed at least four times, followed by two planks to make the soil fine, loose, and level in order to get the best possible germination and development [32]. When Pakistani cotton farmers use a tractor to plow their fields, the earth is torn up to 15 cm deep [33]. In comparison to compacted soil, root hairs often functioned better in uncompacted soil. Due to compacted soil's decreased ability to absorb vital nutrients and water, root development slows down, and root hair function is poor [34]. After rigorous plowing, cotton is successfully sown on ridges [35].

5.2 Fertilization

The amount of fertilizer needed for cotton varies on the soil type, weather, and variety of the crop [36]. For various types of soil, there are different needs. According to a countrywide soil investigation, deficiencies of nitrogen (N), phosphorus (P), and rarely potassium (K) have been noted [37]. It is advised to apply 200 kilograms N, 145 kg P, and 95 kg K per ha to poor soils. N and K dosages for medium and rich soils are the same, while P requirements are 114 and 87 kg/ha, respectively [38]. Due to their early fruiting habits, current early cotton cultivars respond well to the use of N fertilizers. In earlier investigations, cotton cultivars were shown to respond favorably to N and P fertilizers [39].

Taller varieties need less N because of their strong root systems. According to soil testing results, farmers must apply zinc (Zn) and boron (B) at rates of 12 and 9 kg/ha, respectively, in the majority of Pakistan's cotton-growing regions [40]. When sowing, a baseline dosage of the major nutrients like P and K as well as the minor nutrients like Zn and B should be applied [41]. N should be applied in accordance with the crop's requirements and the local climate at the time [42]. In this study, 30 kgN per hectare should be administered once a week for 3 weeks straight in the situation that leaf curl virus symptoms develop [43].

6. Water management

A greater yield is a product of adequate irrigation since it keeps water circulating to the plant's roots and dilutes the vital micro and macronutrients that support healthy plant growth [44]. Salts accumulate in the root zone as a result of insufficient watering, and the plants have little access to nutrients [45]. In contrast, excessive irrigation raises the water table, which pushes salts higher and prevents plants from performing their physiological activities properly since there is insufficient carbon dioxide available to the root zone [46]. The crop's health, the weather, and the sowing technique should all be taken into consideration while scheduling irrigation [47]. A flat-sown crop receives its initial watering 35–40 days after planting, and following irrigations are spaced 12–15 days Apart [48].

By the first 10 days of October, the final irrigation for this type of seeding should have been given [49]. To ensure adequate germination when using the bed planting method, the initial watering should be administered 3–4 days after sowing [50]. Depending on the crop condition and variety, the second and third irrigations should be applied weekly, and the remaining irrigations should be applied every 15 days [51]. To prevent excessive vegetative growth during the wet season, irrigation should be done judiciously [52]. The bed planting method's irrigation termination date is October 15 to allow for the timely sowing of the wheat crop [53].

7. Pest and disease management

One of the main barriers to Pakistani cotton production is insect infestations [54]. The main obstacles to Pakistan's high cotton productivity are sucking insects (including whiteflies, jassids, and aphids) and bollworms (such as the American bollworm, spotted bollworm, and pink bollworm) [55]. The farmers' efforts to manage these insect pests are occasionally hampered by untimely rains and high temperatures, which cause crop losses and significant output decreases [56]. Even with substantial pesticide treatment, farmers were unable to manage the heavy whitefly attack throughout the months of June, July, and August, which led to the loss of cell sap, low boll setting, and low crop output in all cotton-growing regions [57].

Whitefly infestations were severe, which helped CLCuV spread in Punjab's hotspot regions and contributed to poor cotton production [28]. Similar to this, during October's high temperatures and poor Bt variety expression, pink bollworms spread and damaged bolls, resulting in a low cotton yield [58]. Bt (Cry1Ac) gene's low expression after 80 days has reduced its efficiency, making it inappropriate to protect cotton from various bollworms [59]. Although there is currently no scientific evidence, it is possible that some pink bollworm pests with high levels of resistance might migrate from India. The resistant pest may have been transported through the import of Indian cotton and the quarantine department's lack of action [60].

High temperatures during blooming and boll development are yet another major barrier to Pakistan's cotton output [61]. This causes blooms and bolls to fall off, which affects crop yields and output across the nation. Among the cotton diseases, CLCuD caused by several CLCuV (Begomovirus) strains has continued to be a major barrier to Pakistan's ability to produce cotton [62]. The prevalence of this disease is declining in the nation as a result of evolution, where cultivars have a high degree of tolerance against CLCuV [28]. However, in recent years, bacterial cotton blight has started to bring a serious danger to cotton output [63].

Other significant factors contributing to low cotton productivity include poor soil productivity, insufficient fertilizer use, salinity, waterlogging, insufficient water supply, improper soil and water management, pesticide misuse, early or late cotton planting, division between landlords and tenants, low adoption of new technology, financial constraints, and low literacy levels [36, 64].

8. Weed control

Due to weeds' competition for resources including water, light, nutrients, and space, weeds can seriously harm cotton output [65]. Weeds' sluggish appearance and development rates make cotton particularly vulnerable to them [66]. *Amaranthus viridis* L., *Portulaca oleracea* L., *Convolvulus arvensis* L., *Digera arvensis* Forssk., *Xanthium strumariun* L., *Cyperus rotundus L.*, *Echinochloa colona* (L.) Link, and *Cynodon dactylon* (L.) Pers. is among the common cotton weeds in Pakistan [40]. In the first 60 days of crop growth, these weeds are the most damaging because, in addition to contributing to low output owing to the sharing of crop nutrients, they can also act as a secondary home for pests and viruses that affect cotton [67].

Pre-emergence herbicides such pendimethalin or S-metolachlor, which can control the weeds for at least 35–40 days after planting, must be used to control the weeds at the early stage [68]. Post-emergence herbicides including haloxyfop-R-methyl, fluazifop-P-butyl, and quizalofop-p-ethyl have been given permission to be used to manage weeds in cotton [69]. Weeds are eradicated after reemergence either manually or with the help of recently imported Chinese mini-hoeing devices [70]. The majority of farmers combine the use of pesticides with mechanical weed management to effectively control weeds [71]. It is necessary to continue the weeding procedure until the plants form a canopy that shadows the weeds and inhibits their development [72].

9. Criteria for sustainable cotton production

The criteria for sustainable cotton production may vary depending on the certification program or standard being followed. However, here are some common criteria and practices often considered in sustainable cotton production:

1. Environmental Criteria:

   - Reduced Chemical Use: Minimizing the use of synthetic fertilizers, pesticides, and herbicides, and promoting Integrated Pest Management practices.
   - Water Conservation: Efficient irrigation methods, water recycling, and reducing water consumption.
   - Soil Health: Promoting soil fertility, conservation, and erosion control through practices such as crop rotation and cover cropping [73].
   - Biodiversity Conservation: Preserving and enhancing biodiversity on cotton farms by protecting natural habitats and wildlife.

2. Social Criteria:

   - Fair Labor Practices: Ensuring safe and fair working conditions, fair wages, and adherence to labor laws and standards.
   - Workers' Rights: Respect for the rights of workers, including freedom of association, no child or forced labor, and non-discrimination.
   - Community Engagement: Supporting local communities through capacity building, education, healthcare, and infrastructure development.

3. Traceability and Transparency:

   - Documented Supply Chain: Maintaining a transparent and traceable supply chain from farm to final product, allowing for verification and accountability.

- Certification and Auditing: Regular audits and inspections to ensure compliance with sustainability standards and practices.

4. Resource Efficiency:

- Energy Conservation: Promoting energy-efficient practices in farming operations, processing, and transportation.
- Waste Management: Proper handling and disposal of waste, recycling, and reducing waste generation.

5. Continuous Improvement:

- Farmer Training and Support: Providing farmers with education, training, and access to resources to enhance their knowledge and skills in sustainable practices.
- Research and Innovation: Encouraging research and development of sustainable farming techniques and technologies to improve cotton production practices.

It is important to note that the specific criteria and weightage assigned to each criterion may vary among different certification programs. Farmers and cotton industry stakeholders are encouraged to adopt these sustainable practices to reduce the environmental impact of cotton production and promote social responsibility within the supply chain [74].

9.1 To address these challenges, potential solutions include

1. Farmer Support and Capacity Building: Providing training, technical assistance, and access to resources for farmers to adopt sustainable practices and meet certification requirements. This can be achieved through partnerships with local organizations, governments, and industry stakeholders [75].

2. Financial Incentives: Offering financial support, incentives, or premium prices for certified cotton to offset the costs associated with certification and encourage adoption among farmers [76].

3. Collaboration and Partnership: Building strong partnerships and collaborations among stakeholders in the cotton supply chain, including farmers, brands, retailers, and certification bodies. Working together can streamline processes, improve communication, and share knowledge and resources [77].

4. Technology and Data Management: Leveraging digital technologies, such as blockchain and digital platforms, to enhance traceability, data management, and transparency in the cotton supply chain. This can improve supply chain efficiency and facilitate verification processes [78].

5. Consumer Education and Awareness: Investing in consumer education and awareness campaigns to increase understanding of certified cotton and its

benefits. Providing clear and easily recognizable labels and marketing materials that highlight the sustainability credentials of certified cotton can help consumers make informed choices [79].

Addressing these challenges requires a multi-stakeholder approach, with active involvement and collaboration among farmers, industry players, certification bodies, governments, and civil society organizations. By addressing these challenges, cotton certification programs can enhance their effectiveness, increase adoption rates, and drive sustainable practices throughout the cotton supply chain.

10. Case study successful implementation

10.1 Case study: successful integration of sustainable practices in a cotton chain

The textile and clothing (T&C) business plays a crucial role in people's daily lives, and the products it produces are regarded as the second-most important thing that people desire. The sector has grown significantly and achieved great success over the past few decades, making it one of the most significant but also most polluting worldwide industries [80]. Significant environmental and social effects are caused by its magnitude, diversity of processes, and intricate worldwide production networks. Three natural fibers have been used by humans for at least 5000 years each: flax, wool, and cotton.

Today, 48% of all textiles are made from cotton. Only 27% of cotton is grown in rain-fed conditions globally, with irrigated fields accounting for the majority (73%). According to [81], irrigated cotton has an average output of 845 kg per hectare, and rain-fed cotton has an average yield of 391 kg per hectare. 75% of the total global production of cotton is produced by the six biggest cotton-producing nations (China, USA, India, Pakistan, Uzbekistan, and Turkey). Over the past 40 years, the contemporary cotton business has made substantial environmental progress, but it is not letting up.

Researchers and scientists work to create innovative techniques that will enable cotton to be grown, processed and manufactured more effectively and with less environmental impact. The cotton industry will be able to fulfill the demands for productivity and profit while identifying and implementing innovative technologies and practices without sacrificing the ability of future generations to meet their own needs.

10.2 Collaboration among stakeholders

Businesses began signing up for these certification programs to expand their market share and take advantage of fresh opportunities and trends. At the same time, these programs give companies the chance to have more control over their supply chains, ensuring safety and quality requirements while lowering manufacturing costs [82]. Positive side effects of certification included improved reputation and confidence among consumers and other stakeholders.

According to Bolwig *et al.* [83], certification schemes and the methods employed to implement them determine the extent to which these desired beneficial results are realized. In order to enhance the environmental and social circumstances of production operations in developing nations, sustainable chain governance systems have emerged as a result of collaboration between various market and non-market players [84].

10.3 Training

Training farmers helps them become more productive, more able to meet the quality standards demanded by international supply chains, and more capable of producing food.

10.4 Information systems

Giving smallholders access to information and communications can assist them in making decisions and expanding into new or more advantageous markets. According to Parikh *et al.* [85], sound decision-making requires knowledge about the market. Adaptation of sustainable practices should be done at different stages.

Despite the fact that cotton contributes significantly to both national and rural economies, the extensive use of irrigation water, inorganic fertilizer, and pesticides has had a negative impact on the environment, public health, and financial return. Cotton with such significant external inputs and resource requirements is frequently referred to as conventional cotton. A similar innovation occurs in agriculture when a new crop, product, method, or approach to carrying out agricultural tasks is introduced. Innovations that improve sustainability from the perspectives of the economy, society, and environment are referred to as sustainable innovations.

10.5 Land use

In order to build a more sustainable future, agricultural land must be used effectively and efficiently. The good news is that cotton growers have implemented a number of techniques intended to maximize land use effectiveness, such as

1. Improving genetic variety and farmer management to increase cotton fiber production.

2. Encouraging crop rotation, no-till methods, and cover crops to improve soil health.

3. Using more advanced sensors and computer programs for irrigation scheduling.

4. Expanding precise applications through the use of geospatial technologies.

5. Using digital technologies to provide farmers more capacity for prediction.

6. Increasing farmer knowledge of soil health and putting strategies in place that foster a diversified and healthy rhizosphere.

10.6 Pesticides

Thanks to advancements in biotechnology and pest management, cotton growers may use a lot less pesticides and insecticides now than they did in the past. In actuality, cotton consumes an average of only five grams of pesticides per kilogram compared to other crops. The Boll Weevil Eradication Program, started in the 1970s, was one of the most successful uses of Integrated Pest Management in history. Boll weevils used to cost American farmers $300 million in crop damage per year; however, all states in the union that grow cotton have eradicated the pest entirely, with the exception of Texas.

Thanks to this innovative method, cotton growers have been able to cut their insecticide applications by 50 percent since the 1980s. Many of those that are still in use target specific species, protecting other organisms from damage and enabling them to continue existing and contributing to the ecosystem. Sustainable supply chain management, which coordinates demand needs toward suppliers along the supply chain while addressing economic as well as social and environmental issues, is strongly emphasized by corporate responsibility in global sourcing operations [86].

10.7 Positive impacts on environment and social aspects

Cotton is the most profitable and extensively cultivated non-food crop globally. Its manufacture generates income for over 250 million people and employs more than 7% of the labor force in underdeveloped countries. Approximately half of all textiles are made of cotton. Despite cotton's enormous worldwide market, the unsustainable nature of the current production techniques threatens the industry's capacity to continue producing the crop in the future. Bringing cotton production into compliance with even the most minimally acceptable environmental standards is a challenging task. To encourage the ethical production and use of cotton, WWF is working with a variety of international partners in various capacities. The three main environmental impacts of cotton are water use, habitat modification, and the use of agrochemicals, especially pesticides.

10.8 Water scarcity

Water is extensively used in the growth and processing of cotton. According to some experts, among all agricultural products, cotton uses the most water. To irrigate cotton crops, surface and ground fluids are frequently diverted, resulting in freshwater loss through evaporation and ineffective water management.

10.9 Soil erosion and degradation

Water is used a lot in the growth and processing of cotton. Out of all agricultural products, cotton is said to require the most water by certain specialists. Frequent diversion of surface and groundwater for cotton crop irrigation results in freshwater loss due to evaporation and ineffective water management.

10.10 Pollution

Conventional cotton farming practices involve heavy use of pesticides and fertilizers. Pesticides endanger not only the water and soil quality but also the biodiversity in and around the fields. Because pesticides are used so extensively, there is also concern for the health of farm workers and the neighboring populations.

10.11 Water contamination

Pesticides, fertilizers, and minerals are introduced into rivers, lakes, marshes, and subsurface aquifers by cotton field runoff. Because of their rapid toxicity, these pollutants either directly or indirectly destroy biodiversity through long-term accumulation.

11. Future directions and conclusions

11.1 Technological developments for the sustainable production of cotton

The implementation of these key technologies is required to attain sustainable cotton production.

11.1.1 Single-seed precision sowing technology

High-quality finishing seeds, meticulous soil preparation, a sensible row spacing arrangement, mechanical seeding, and no thinning or final singling of seedlings after emergence are all part of the single-seed precision sowing method for cotton. High-quality finishing seeds, meticulous soil preparation, a sensible row spacing arrangement, mechanical seeding, and no thinning or final singling of seedlings after emergence are all part of the single-seed precision sowing method for cotton. This method also does away with the need for planting, thinning, and final singling.

11.1.2 Light and simplified seedling nursery technology

A novel method called light and simple nursery and transplanting of cotton seedlings is intended to replace the conventional matrix and water floating seedling nurseries, as well as traditional nutrition bowl nursery and transplanting. The general protocol of conventional nutrition bowl nurseries and transplanting is followed by these nursery techniques. To create an industrial nursery, they replace ordinary soil with commercial matrices (combinations of peat, vermiculite, and river sand) or nutrient solutions combined with chemicals that promote root development, preserve leaves, and control other aspects of plant growth. The number of seedlings per unit area rises as a result of the alternative to the common soil, and cotton seedlings with little or no soil are utilized for transplanting. Additionally, the use of machinery for cotton seedling transplantation guarantees mechanized transplanting, lowers labor intensity, and increases productivity. In light of this, the new technology for cotton seedling nursery and transplanting is somewhat lighter and simpler than the conventional methods.

11.1.3 One-time fertilization technique

The three main cotton-growing regions of China rely on light, easy, and effective fertilization, with the amount of fertilizer used determined by the output of the cotton. The appropriate amount is 225–240 kg/ha for high-yield fields with an area of 3750 kg/ha or higher, and the $N:P_2O_5:K_2O$ ratio is roughly 1.0:0.6:0.7–0.9. The yield of seed cotton in the northwest inland cotton area is between 4500 and 5250 kg/ha, and 280 to 330 kg/ha of N is the appropriate amount. The $N:P_2O_5:K_2O$ ratio is roughly 1.0 0.5:0-0.3, and it can be decreased by 15% by integrating water fertilizer. The other is lowering the frequency of fertilization by applying cotton special slow-release fertilizer. The Yangtze River Valley cotton area uses a one-time basal application of controlled or slow-release fertilizer; 10–20% less slow-release fertilizer can be equivalent to the standard fertilization dosage.

11.1.4 Water and fertilizer integration technology

Optimizing water efficiency during irrigation is the goal of cotton water-saving irrigation technology. The most popular water-saving irrigation system that combines surface drip irrigation and plastic film mulching is drip irrigation under plastic mulching, which is extensively utilized in the northwest inland region. It supplies water via a low-pressure pipeline system, filters the pressurized water using filter facilities, and dissolves the fertilizer in water to create an aqueous solution. To keep the concentrated cotton root regions at the proper moisture content for the incorporation of water and fertilizer, this solution permeates them uniformly. The average water consumption for sprinkler irrigation is 50%, compared to 12% for traditional irrigation. The amount of fertilizer needed is lowered by 15–20%.

12. Research and development priorities

Under the guidance of PCCC, three cotton research institutions and seven research stations collaborate to improve cotton output. Several experimental research and development initiatives aid in identifying bottleneck problems for the production of sustainable cotton. Multidisciplinary research at the Central Cotton Research Institutes (CCRI) at Sakrand, Sindh, and Multan, Punjab, has greatly aided in the development of improved cultivars. Moreover, CCRI Multan controls germplasm resource distribution, identification, evaluation, maintenance, and preservation. Moreover, the primary government institutions that carry out research on numerous cotton-related themes are Bahauddin Zakariya University (BZU), Islamia University Bahawalpur (IUB), University of Agriculture Faisalabad (UAF), and Mohammad Nawaz Shareef University of Agriculture, Multan (MNS-UAM). Through the planning of several workshops, training sessions, and seminars, these organizations support the growing of cotton. Cotton breeding initiatives have been started by numerous recognized national and worldwide private commercial business companies in addition to state organizations. Their contribution to cotton research and development has been impaired, meanwhile, by a shortage of skilled breeding personnel, ineffective resources, and outdated breeding facilities.

12.1 Breeding for stress resilient cotton in Pakistan

a. Stress hazards, which are often classified as either biotic or abiotic, have a significant impact on cotton output. Abiotic stresses include heat, dryness, and salinity are examples of environmental conditions, whereas biotic stresses include pests, viruses, and fungal diseases. Plant growth was altered morphologically, physiologically, and biochemically as a result of the extended stress. Pakistan is a country where cotton is very susceptible to a variety of biotic and abiotic stressors, which combined reduce cotton yield and fiber quality. The yield per hectare and total cotton production have recently been negatively impacted by sucking insects, bollworms, cotton leaf curl virus, and extreme temperatures.

b. Biotic stress-resilient cotton: The production of cotton in Pakistan is severely declined due to attack of many chewing and sucking insect pest. Even after multiple Bt cultivars were introduced, the resistant cultivars are still being evaluated and are in the development stage. The cotton plant sustained major damage as a result of the severe bollworm invasion throughout the growth season. The main cause of the rise in pink bollworm in recent years is higher temperatures throughout September and October. The army bollworm's extreme invasion has a significant impact on cotton output. However, army worms' efficient management methods are still inadequate to eradicate huge populations at once. Therefore, creating resistance cultivars is the only practical way to manage bollworms and create Integrated Pest Management (IPM) plans. The cultivars CIM-499, CRIS-468, MNH-633, MNH-635, and MNH-886 have all been approved. Targeted genome editing may therefore be essential to creating cotton that is resistant to bollworms.

c. In cotton, sucking insects can diminish production by roughly 30–50%. In both Bt and non-Bt cotton cultivars, the mealy bug, dusky bug, jassid, and whitefly seemed to be the most common detrimental pests that drain the sap of the plant during the vegetative and reproductive stages. Various breeding strategies were employed to create cultivars of CLCuD resistance. During the first outbreak, the resistance of NIAB-846, NIAB-111, CIM-443, Bt-Cyto-178, Cyto-124, MNH-886, and FH-142 was better.

d. Abiotic stress-resilient cotton: Global warming is causing harm to the world's sustainable cotton production. Cotton production is adversely affected by abiotic conditions such as heat, drought, salt, and nitrogen shortage. In Pakistan, the cotton-producing season is typically characterized by high temperatures between 40 and 45 degrees Celsius. The cotton crop's potential output is limited by low or high temperatures. Cold temperatures during the seeding stage cause delayed germination and uneven crop establishment. [87]. Furthermore, the physiological, metabolic, and biochemical processes of boll formation, lint output, and fiber quality were all impacted by high-temperature stress. Prior research in breeding demonstrated the value of using bulk selection in conjunction with pedigree selection to produce heat-tolerant cultivars. The CIM-240, CIM-598, CIM-777, NIAB-999, NIAB-111, and CIM-602 exhibit superior adaptability to heat stress. Cotton is a moderately salt-tolerant plant, but the level of tolerance differs among genotypes [88]. It is more vulnerable to salt stress in the early stages of seedling emergence and germination. High salt content prevents the growth of roots and shoots, which eventually alters cotton output and fiber quality [89]. NIAB-78, FH-113, MNH-93, NIAB-999, and CIM-707 all showed improved salt tolerance when cultivated. However, the primary breeding problem facing the majority of Pakistani cotton producers remains the development of cultivars resistant to salt. To improve cotton output per acre, critical nutrients must be available. Cotton's growth was hampered by a substantial nutritional deficit. On the other hand, overnutrition has harmful consequences. Breeding in Pakistan is still restricted when it comes to genotypes that are highly input-responsive.

12.2 Marker-assisted breeding

- The effectiveness of traditional plant breeding operations has greatly increased with the latest developments in molecular markers. Linkage maps are frequently created using various molecular markers in order to identify quantitative trait loci (QTLs). Cotton markers include:

- Simple sequence repeats (SSRs).
- Single nucleotide polymorphism (SNP).
- Amplified fragment length polymorphism (AFLP).
- Random Amplified Polymorphic DNA (RAPD).

shown to be beneficial for investigating genetic diversity, genetic mapping of desired phenotypes, determining candidate gene nucleotide sequences and assisting with marker-assisted breeding (MAS). There has been significant progress in identifying QTLs related to agronomic traits, yield components, fiber quality, plant architecture, leaf trichomes, leaf pubescence, seed oil, seed gossypol contents, early maturity, heat tolerance, drought tolerance, salt tolerance, and disease resistance thanks to QTL mapping using a suitable cotton mapping population. In Pakistan, NIBGE-2 and NIBGE-115 were bred using a combination of traditional methods and DNA markers linked to resistance to the cotton leaf curl viral disease. Likewise, MAS was used to produce the resistance genotypes CIM-443 and CIM-240 against the cotton leaf curl viral disease. The CRIS-134 cultivar was found to have a variety of QTLs for drought tolerance when MAS was applied to it. In the meantime, Pakistan's widespread application of molecular markers for MAS is still in its infancy.

13. Conclusion and key take away

About 1.7 million people in Pakistan are involved in the value chain of cotton farming. Pakistan ranks fourth globally in terms of the amount of cotton lint produced. Nowadays, a multitude of biotic and abiotic variables provide numerous obstacles for cotton. Biotic variables, which are major contributors to the decline in cotton yield, include weeds, illnesses, and insect pests. Weeds and pest insects that feed on nectar are major problems in the beginning. Using pesticides is not the only approach to managing insect infestations. Starting with the selection of seed kinds resistant to pest insects is the right course of action. Every element, including fertilizer use and irrigation, contributes to the proliferation of pests. The attack of sucking insect pests is increased when nitrogenous fertilizers are used excessively. It is advisable to promote the use of green lacewing and ladybird beetles to reduce the number of sucking pests, such as whiteflies and jassids. Various management techniques are used depending on the type of insect problem.

Author details

Mirza Abdul Qayyum[1*], Shafqat Saeed[1], Waqas Wakil[2], Muhammad Yasin[3], Ahmad Nawaz[4], Abid Ali[2], Habib Ali[5], Unsar Naeem-Ullah[1], Umer Sharif[1], Hasan Taha[1], Aiman Khalid[1], Iqra Maryum[1], Zeeshan Shahid[1] and Wania Hasan Zubairi[1]

1 Institute of Plant Protection, MNS University of Agriculture, Multan, Pakistan

2 Department of Entomology, University of Agriculture, Faisalabad, Pakistan

3 Department of Entomology, Islamia University Bahawalpur, Pakistan

4 Department of Plant Sciences, College of Agricultural and Marine Sciences, Sultan Qaboos University, Muscat, Oman

5 Department of Entomology, Khwaja Fareed University of Engineering and Information Technology, Rahim Yar Khan, Pakistan

*Address all correspondence to: qayyum.mirza@mnsuam.edu.pk

References

[1] Abbas HG, Mahmood A, Ali Q. Zero tillage: A potential technology to improve cotton yield. Genetika. 2016;**48**(2):761-776

[2] Goldbach M, Seuring S, Back S. Co-ordinating sustainable cotton chains for the mass market. Greener Management International. 2003;**43**:65-78

[3] Zia SM, Inayatullah C, Samee M, Hasnain T. Sustainable Cotton Production, Trade and Environmental Impact: Policy Issues and Options for Pakistan. Pakistan: Sustainable Development Policy Institute; 1998

[4] Brader L. Integrated pest control in the developing world. Annual Review of Entomology. 1979;**24**(1):225-254

[5] Chapagain AK, Hoekstra A, Savenije H, Gautam R. The water footprint of cotton consumption: An assessment of the impact of worldwide consumption of cotton products on the water resources in the cotton producing countries. Ecological Economics. 2006;**60**(1):186-203

[6] Josipović M, Plavšić H, Kovačević V, Marković M, Iljkić D. Impacts of irrigation and genotype on yield, protein, starch and oil contents in grain of maize inbred lines. Genetika. 2014;**46**(1):243-253

[7] Khaliq A, Abbasi MK, Hussain T. Effects of integrated use of organic and inorganic nutrient sources with effective microorganisms (EM) on seed cotton yield in Pakistan. Bioresource Technology. 2006;**97**(8):967-972

[8] Myers D, Stolton S. Organic Cotton: From Field to Final Product. London, U.K: Intermediate Technology Publications Ltd (ITP); 1999

[9] Luz C. Waste couture: Environmental impact of the clothing industry. Environmental Health Perspectives. 2007;**115**(9):A448

[10] Muthu SS. Roadmap to sustainable textiles and clothing: Eco-friendly raw materials, technologies, and processing methods. New York, Dordrecht, London: Springer; 2014

[11] Elahi M, Cheema S, Basra MA, Ali Q. Use of Allelopathic water extract of field crops for weed control in wheat. International Research of Journal of Plant Science. 2011;**2**(9):262-270

[12] Husnain T. Transformation and transgenic expression studies of glyphosate tolerant and cane borer resistance genes in sugarcane (*Saccharum officinarum* L.). Molecular Plant Breeding. 2015;**6**(12):1-17

[13] Singh K, Singh H, Singh K. Weed management in cotton with pre-and post-emergence herbicides. Indian Journal of Weed Science. 2016;**48**:348. DOI: 10.5958/0974-8164.2016.00090.3

[14] Choudhury AR. Development of Ecolabels for Sustainable Textiles: Roadmap to Sustainable Textiles and Clothing. Springer; 2015. pp. 137-173

[15] Ali A, Hussain T, Tantashutikun N, Hussain N, Cocetta G. Application of smart techniques, internet of things and data mining for resource use efficient and sustainable crop production. Agriculture. 2023;**13**(2):397

[16] Ali H, Hussain GS, Hussain S, Shahzad AN, Ahmad S, Ahmad HMR, Sarwar N. Early sowing reduces cotton leaf curl virus occurrence and improves cotton productivity.

Agronomical Research in Moldavia/ Cercetări Agronomice în Moldova. 2014;**60**(4):71-81

[17] Hou QQ, Chen BM, Peng SL, Chen LY. Effects of extreme temperature on seedling establishment of nonnative invasive plants. Biological Invasions. 2014;**16**:2049-2061

[18] Sufyan Tahir M, Latif A, Bashir S, Shad M, Khan MAU, Gul A, et al. Transformation and evaluation of broad-spectrum insect and weedicide resistant genes in *Gossypium arboreum* (Desi Cotton). GM Crops & Food. 2021;**12**(1):292-302

[19] Karaca M, Ince AG. A DNA extraction method for nondestructive testing and evaluation of cotton seeds (*Gossypium* L.). Biochemical Genetics. 2023. DOI: 10.1007/s10528-023-10496-5

[20] Antneh DN. Effect of dosage and duration of sulfuric acid on delinting, seed germination and seedling vigor of cotton (*Gossypium hirsutum L.*) [Doctoral dissertation]; Department of Plant Sciences, College of Agriculture and Rural Transformation, University of Gondar); 2017

[21] Xu B, Gao M, Hu W, Zhao W, Wang S, Zhou Z. Individual and interactive influences of elevated air temperature and soil drought at the flowering and boll-forming stage on cottonseed yield and nutritional quality. The Crop Journal. 2022;**10**(1):128-139

[22] Lima JME, Carvalho ER, Moraes LFDS, Cossa NHDS, Miquicene FVC, Gradela YF. Delinting and neutralizers residue effect on stored cotton seeds physiological quality determined by phenotyping image analysis. Journal of Seed Science. 2023;**45**:e202345014

[23] Paz JO, Woli P, Garcia AGY, Hoogenboom G. Cotton yields as influenced by ENSO at different planting dates and spatial aggregation levels. Agricultural Systems. 2012;**111**:45-52

[24] Mudassir MA, Rasul F, Khaliq T, Yaseen M. Conformance of sowing dates for maximizing heat use efficiency and seed cotton yield in arid to semi-arid cotton zone of Pakistan. Environmental Science and Pollution Research. 2021;**19**:1-15

[25] Dong H, Li W, Tang W, Zhang D. Furrow seeding with plastic mulching increases stand establishment and lint yield of cotton in a saline field. Agronomy Journal. 2008;**100**(6):1640-1646

[26] Hollinger SE, Angel JR. Weather and crops. In: Illinois Agronomy Handbook. 2009. pp. 1-12

[27] Fernando M, Shrestha A. The potential of cover crops for weed management: A sole tool or component of an integrated weed management system? Plants. 2023;**12**(4):752

[28] Farooq J, Farooq A, Riaz M, Shahid MR, Saeed F, Iqbal MS, et al. Cotton leaf curl virus disease a principle cause of decline in cotton productivity in Pakistan (a mini review). Canadian Journal of Plant Protection. 2014;**2**:9-16

[29] Constable GA, Bange MP. The yield potential of cotton (*Gossypium hirsutum* L.). Field Crops Research. 2015;**182**:98-106

[30] McDonald MF, Copeland LO. Seed Production: Principles and Practices. Springer Science & Business Media; 2012

[31] Katsvairo TW, Wright DL, Marois JJ, Hartzog DL, Balkcom KB, Wiatrak PP,

et al. Cotton roots, earthworms, and infiltration characteristics in sod–peanut–cotton cropping systems. Agronomy Journal. 2007;**99**(2):390-398

[32] Singh K, Mishra SK, Singh M, Singh K, Brar AS. Water footprint assessment of surface and subsurface drip fertigated cotton-wheat cropping system–a case study under semi-arid environments of Indian Punjab. Journal of Cleaner Production. 2022;**365**:132735

[33] Raza W, Yousaf S, Niaz A, Rasheed MK, Hussain I. Subsoil compaction effects on soil properties, nutrient uptake and yield of maize fodder (*Zea mays* L.). Pakistan Journal of Botany. 2005;**37**(4):933

[34] Haling RE, Brown LK, Bengough AG, Young IM, Hallett PD, White PJ, et al. Root hairs improve root penetration, root-soil contact, and phosphorus acquisition in soils of different strength. Journal of Experimental Botany. 2013;**64**(12):3711-3721

[35] Accorsi E. Tape your way to sustainability: A Life Cycle Assessment of a sport tape made from conventional cotton and ways to minimize its environmental impact [Doctoral dissertation]; Politecnico di Torino; 2023)

[36] Khan A, Tan DKY, Afridi MZ, Luo H, Tung SA, Ajab M, et al. Nitrogen fertility and abiotic stresses management in cotton crop: A review. Environmental Science and Pollution Research. 2017;**24**:14551-14566

[37] Ryan J, Rashid A, Torrent J, Yau SK, Ibrikci H, Sommer R, et al. Micronutrient constraints to crop production in the Middle East–West Asia region: significance, research, and management. Advances in Agronomy. 2013;**122**:1-84

[38] Ullah A, Ahmad I, et al. Optimizing management options through empirical modeling to improve pearl millet production for semi-arid and arid regions of Punjab, Pakistan. Sustainability. 2020;**12**(18):7715

[39] Shahzad AN, Rizwan M, Asghar MG, Qureshi MK, Bukhari SAH, Kiran A, et al. Early maturing Bt cotton requires more potassium fertilizer under water deficiency to augment seed-cotton yield but not lint quality. Scientific Reports. 2019;**9**(1):7378

[40] Ali MA, Farooq J, Batool A, Zahoor A, Azeem F, Mahmood A, et al. Cotton production in Pakistan. Cotton Production. 2019:249-276

[41] Seleiman MF, Al-Suhaibani N, El-Hendawy S, Abdella K, Alotaibi M, Alderfasi A. Impacts of long-and short-term of irrigation with treated wastewater and synthetic fertilizers on the growth, biomass, heavy metal content, and energy traits of three potential bioenergy crops in arid regions. Energies. 2021;**14**(11):3037

[42] Rigby H, Clarke BO, Pritchard DL, Meehan B, Beshah F, Smith SR, et al. A critical review of nitrogen mineralization in biosolids-amended soil, the associated fertilizer value for crop production and potential for emissions to the environment. Science of the Total Environment. 2016;**541**:1310-1338

[43] Dingus A. Characterizing Fonio production across scales: Interdisciplinary insights from a local field trial and a scoping review [Doctoral dissertation]; Michigan State University. 2022

[44] Roy RN, Finck A, Blair GJ, Tandon HLS. Plant nutrition for food security. A guide for integrated nutrient management. FAO Fertilizer and Plant Nutrition Bulletin. 2006;**16**(368)

[45] Li C, Lei J, Zhao Y, Xu X, Li S. Effect of saline water irrigation on soil development and plant growth in the Taklimakan Desert Highway shelterbelt. Soil and Tillage Research. 2015;**146**:99-107

[46] Wassmann R, Jagadish SVK, Heuer S, Ismail A, Redona E, Serraj R, et al. Climate change affecting rice production: The physiological and agronomic basis for possible adaptation strategies. Advances in Agronomy. 2009;**101**:59-122

[47] Chapagain T, Yamaji E. The effects of irrigation method, age of seedling and spacing on crop performance, productivity and water-wise rice production in Japan. Paddy and Water Environment. 2010;**8**:81-90

[48] Das TK, Bhattacharyya R, Sudhishri S, Sharma AR, Saharawat YS, Bandyopadhyay KK, et al. Conservation agriculture in an irrigated cotton–wheat system of the western Indo-Gangetic Plains: Crop and water productivity and economic profitability. Field Crops Research. 2014;**158**:24-33

[49] Fischer RA. The effect of duration of the vegetative phase in irrigated semi-dwarf spring wheat on phenology, growth and potential yield across sowing dates at low latitude. Field Crops Research. 2016;**198**:188-199

[50] Gopal R, Jat RK, Malik RK, Kumar V, Alam MM, Jat ML, et al. Direct dry seeded rice production technology and weed management in rice based systems. Gates Open Research. 2019;**3**(207):207

[51] Gutiérrez J, Villa-Medina JF, Nieto-Garibay A, Porta-Gándara MÁ. Automated irrigation system using a wireless sensor network and GPRS module. IEEE Transactions on Instrumentation and Measurement. 2013;**63**(1):166-176

[52] Costa JM, Ortuño MF, Chaves MM. Deficit irrigation as a strategy to save water: Physiology and potential application to horticulture. Journal of Integrative Plant Biology. 2007;**49**(10):1421-1434

[53] Kajla M, Yadav VK, Chhokar RS, Sharma RK. Management practices to mitigate the impact of high temperature on wheat. Journal of Wheat Research. 2015;**7**(1):1-12

[54] Nagrare VS, Kranthi S, Biradar VK, Zade NN, Sangode V, Kakde G, et al. Widespread infestation of the exotic mealybug species, *Phenacoccus solenopsis* (Tinsley) (Hemiptera: Pseudococcidae), on cotton in India. Bulletin of Entomological Research. 2009;**99**(5):537-541

[55] Javed K, Jabbar A, Bhatt S, Mahmood K, Javed B. A comparative study of various controlling methods to monitor pink bollworm (*Pectinophora Gossypiella*) infestation on the yield of cotton: A case study of district Vehari, Pakistan. Big Data in Agriculture (BDA). 2020;**2**(1):29-31

[56] Hill DS. Pests of Crops in Warmer Climates and their Control. Springer Science & Business Media; 2008

[57] Conboy NJ, McDaniel T, George D, Ormerod A, Edwards M, Donohoe P, et al. Volatile organic compounds as insect repellents and plant elicitors: An integrated pest management (IPM) strategy for glasshouse whitefly (*Trialeurodes vaporariorum*). Journal of Chemical Ecology. 2020;**46**:1090-1104

[58] Mohan KS, Ravi KC, Suresh PJ, Sumerford D, Head GP. Field resistance to the *Bacillus thuringiensis* protein

Cry1Ac expressed in Bollgard® hybrid cotton in pink bollworm, *Pectinophora gossypiella* (Saunders), populations in India. Pest Management Science. 2016;**72**(4):738-746

[59] Dong HZ, Li WJ. Variability of endotoxin expression in Bt transgenic cotton. Journal of Agronomy and Crop Science. 2007;**193**(1):21-29

[60] Dent D, Binks RH. Insect Pest Management. Cabi; 2020

[61] Zafar MM, Zhang Y, Farooq MA, Ali A, Firdous H, Haseeb M, et al. Biochemical and associated agronomic traits in *Gossypium hirsutum* L. under high temperature stress. Agronomy. 2022;**12**(6):1310

[62] Brown JK, Khan Z. Breeding cotton for cotton leaf curl disease resistance. In: Cotton Breeding and Biotechnology. Boca Raton, Florida, United States: CRC Press; 2022. pp. 171-197

[63] Priya R, Ramesh D, Khosla E. Biodegradation of pesticides using density-based clustering on cotton crop affected by *Xanthomonas malvacearum*. Environment, Development and Sustainability. 2020;**22**:1353-1369

[64] Sattar T. A sociological analysis of constraining factors of development in agriculture sector of Pakistan. Journal of Economics and Sustainable Development. 2012;**3**(8):8-24

[65] Kaur S, Kaur R, Chauhan BS. Understanding crop-weed-fertilizer-water interactions and their implications for weed management in agricultural systems. Crop Protection. 2018;**103**:65-72

[66] Heap I, Duke SO. Overview of glyphosate-resistant weeds worldwide. Pest Management Science. 2018;**74**(5):1040-1049

[67] Yuan D, Bassie L, Sabalza M, Miralpeix B, Dashevskaya S, Farre G, et al. The potential impact of plant biotechnology on the millennium development goals. Plant Cell Reports. 2011;**30**:249-265

[68] Iqbal N, Manalil S, Chauhan BS, Adkins SW. Investigation of alternate herbicides for effective weed management in glyphosate-tolerant cotton. Archives of Agronomy and Soil Science. 2019;**65**(13):1885-1899

[69] Zhu J, Wang J, DiTommaso A, Zhang C, Zheng G, Liang W, et al. Weed research status, challenges, and opportunities in China. Crop Protection. 2020;**134**:104449

[70] Rana SS, Rana MC. Principles and Practices of Weed Management. Palampur: Department of Agronomy, College of Agriculture, CSK Himachal Pradesh Krishi Vishvavidyalaya; 2016. p. 138

[71] Chauhan BS, Ahmed S, Awan TH, Jabran K, Manalil S. Integrated weed management approach to improve weed control efficiencies for sustainable rice production in dry-seeded systems. Crop Protection. 2015;**71**:19-24

[72] Duke SO, Gutterman Y. Flowering, seed development, and the influences during seed maturation on seed germination of annual weeds. In: Weed Physiology. Boca Raton, Florida, United States: CRC Press; 2018. pp. 1-25

[73] Bergtold JS, Ramsey S, Maddy L, Williams JR. A review of economic considerations for cover crops as a conservation practice. Renewable Agricultural Food System. 2019;**34**:62-76

[74] Lund-Thomsen P, Riisgaard L, Coe NM, Singh S, Ghori S. Bridging

Global Standard Requirements and Local Famer Needs: Implementing Parterns of the Better Cotton Initiative in Pakistan and India. Copenhagen Business School [wp]. CBDS Working Paper No. 1, 2018

[75] Lee JSH, Rist L, Obidzinski K, Ghazoul J, Koh LP. No farmer left behind in sustainable biofuel production. Biological Conservation. 2011;**144**:2512-2516

[76] Addis T, Kachi A, Wang J. A review of current state and future directions of cotton production in Ethiopia. Cogent Food & Agriculture. 2021;7:1880533

[77] Bitzer V, Glasbergen P. Business–NGO partnerships in global value chains: Part of the solution or part of the problem of sustainable change? Current Opinion in Environment Sustainability. 2015;**12**:35-40

[78] Hardin RG IV, Barnes EM, Delhom CD, Wanjura JD, Ward JK. Internet of things: Cotton harvesting and processing. Computers and Electronics in Agriculture. 2022;**202**:107294

[79] Martins SB, Vascouto V. Challenges to present fashion consuming society and market possibilities of organic cotton: A sustainable proposition. In: International Symposium on Sustainable Design, UFPR, Curitiba. Citeseer; 2007

[80] Allwood JM, Laursen SE, de Rodriguez CM, Bocken NM. Well dressed?: The present and future sustainability of clothing and textiles in the United Kingdom. Journal of the Home Economics Institute of Australia. 2015;**22**(1):42

[81] Amos CC, Rahman A, Gathenya JM, Friedler E, Karim F, Renzaho A. Roof-harvested rainwater use in household agriculture: Contributions to the sustainable development goals. Water. 2020;**12**(2):332

[82] Raynolds LT, Murray D, Heller A. Regulating sustainability in the coffee sector: A comparative analysis of third-party environmental and social certification initiatives. Agriculture and Human Values. 2007;**24**(2):147-163

[83] Bolwig S, Gibbon P, Jones S. The economics of smallholder organic contract farming in tropical Africa. World Development. 2009;**37**(6):1094-1104

[84] Vermeulen W, Uitenboogaart Y, Pesqueira LD, Metselaar J, Kok M. Roles of governments in multi-actor sustainable supply chain. Governance Systems and the Effectiveness of Their Interventions. An Exploratory Study. 2010

[85] Parikh TS, Patel N, Schwartzman Y. A survey of information systems reaching small producers in global agricultural value chains. In: 2007 International Conference on Information and Communication Technologies and Development. Piscataway, New Jersey, United States: IEEE; 2007. pp. 1-11

[86] Seuring S, Müller M. Core issues in sustainable supply chain management – A Delphi study. Business Strategy and the Environment. 2007;**2**(7):1-13

[87] Ahmad F, Perveen A, Mohammad N, Ali MA, Akhtar MN, Shahzad K, et al. Heat stress in cotton: Responses and adaptive mechanisms. In: Cotton Production and Uses: Agronomy, Crop Protection, and Postharvest Technologies. Springer Nature Singapore Pvt. Ltd; 2020. pp. 393-428

[88] Wang N, Qi H, Qiao W, Shi J, Xu Q, Zhou H, et al. Cotton (*Gossypium hirsutum* L.) genotypes with contrasting K+/Na+ ion homeostasis: Implications for salinity tolerance. Acta Physiologiae Plantarum. 2017;**39**(3):77

[89] Sharif I, Aleem S, Farooq J, Rizwan M, Younas A, Sarwar G, et al. Salinity stress in cotton: Effects, mechanism of tolerance and its management strategies. Physiology and Molecular Biology of Plants. 2019;**25**:807-820

Section 3

Sustainability and Precision Agriculture in Cotton Production

Chapter 6

A Review of the Factors Affecting Adoption of Precision Agriculture Applications in Cotton Production

Songül Gürsoy

Abstract

Precision agriculture (PA) is a modern farming management system adopted throughout the world, which employs cropping practices by observing and measuring the temporal and spatial variability in fields to enhance the sustainability of agricultural production through more efficient use of land, water, fuel, fertilizer, and pesticides. The efficiency of precision agriculture technologies (PAT) in agricultural production mainly depends on the use of site-specific agricultural inputs accurately through decision support mechanisms by observing and measuring the variables such as soil condition, plant health, and weed intensity. Although there have been significant developments in PAT, especially remote sensing as a key source of information available in support of PA in recent years, its adoption has been very slow by farmers due to a variety of reasons. The main aim of this chapter is to provide a critical overview of how recent developments in sensing technologies, geostatistical analysis, data fusion, and interpolation techniques can be used in the cotton production systems to optimize yields while minimizing water, chemical pesticide, and nitrogen inputs and analysis the main factors influencing the adoption of PAT by cotton farmers. Therefore, this chapter includes a compressive literature survey of the studies done on the current use and trends of PAT, and on farm level use of PA in cotton production worldwide.

Keywords: precision agriculture, cotton management, farmers' adoption, unmanned aerial system, variable rate application

1. Introduction

Cotton is a very important plant throughout the world. The amount of cotton production in the world is nearly 25 million tons of cotton. Leading cotton-producing countries worldwide in 2022/2023 are announced as China, India, United States, and Brazil. It is estimated that world cotton production will reach 28 million tons with an annual increase of 1.5% until 2030 [1, 2]. While cotton production has steadily increased over the past few years, many issues have also started to emerge with the increase in cotton farming areas because current cotton production methods are not environmentally sustainable. Intensive and incorrect use of technological inputs such as tillage, fertilizers, irrigation, pesticides, and herbicides in cotton agriculture has significantly caused soil and environmental degradation as well as reduced crop profitability [3].

One of the new ways that modern agriculture could potentially maintain or enhance crop yields by minimizing environmental pollution is site-specific application of inputs according to the needs of the crop, which is defined as Precision agriculture (PA) [4]. PA is an umbrella term for using modern data-driven technologies to optimize crop management and improve productivity, efficiency, and sustainability in agricultural production. Therefore, PA can be defined as the application of modern information technologies such as GPS, sensors, drones, Internet of Things (IoT), artificial intelligence (AI), and data analytics in the management of crop production [5]. It is seen that studies on PA have gained importance in recent years. The fact that Internet of Things (IoT), artificial intelligence (AI), remote sensing, and image processing (ImP) techniques have been actively used in agriculture by integrating with geographic information systems (GIS) and geographic position systems (GPS) has brought about important developments in the use of precision agriculture technologies (PAT) in agricultural production [6–8]. Kırkaya [6] stated that in the future, PAT will be widespread used in crop management practices such as sowing, fertilization, irrigation, and weed control. The author emphasized that PAT had the ability to protect crop health, soil, and the environment by effective and optimized application of inputs.

PAT reduces not only production costs and increases income for the producer but also reduces the negative environmental impact of agricultural chemicals by adjusting input application rates to crop requirements because it can help farmers monitor and control various aspects of their fields, such as soil conditions, crop growth, pest infestation, and water use [9, 10]. However, PA also has some limitations and challenges such as high initial investment and maintenance costs because it needs expensive and complex equipment, such as sensors, drones, satellites, computers, and software, to collect and analyze data and control the farming operations [11–13]. Therefore, it seems that the PA is not applicable especially in developing countries due to the presence of poor farmers, subsistence farming systems, small farmlands, lack of technical and software knowledge among farmers, and the high cost of application of its technologies.

In cotton agriculture, the adoption of PA technology has been very different than in grain production since cotton needs intensive management processes such as multiple fertilizer applications, multiple plant growth regulator applications, multiple irrigation potential applications, and multiple pesticide applications. The availability of cotton yield sensors later than grain yield sensors affected cotton producers' adaptation to precision agriculture.

In this chapter, a review of PA techniques and practices used in cotton production is presented along with several considerations and challenges. The advantages and disadvantages of PA, as well as some of the current and future trends and opportunities for the usage of PAT in crop management practices in cotton production are explored.

2. Precision agriculture and its practices in cotton production

Cotton agriculture is known for its intensive crop management practices with higher levels of input including seeds, land management practices, agrochemicals (fertilizers, pesticides, and herbicides), and water, which is negatively influencing farmers' profit and yield [14]. Also, cotton crop is host to many insects and pests, and enormous amounts of pesticides are used which are hazardous for the environment in the long run, in order to control these insects and pests [15]. When PAT is applied to these management practices, it will be possible to improve the economic and environmental sustainability of cotton production.

Several factors, such as the differences in soil texture and fertility and the occurrence of pests, diseases, and nematodes, can cause spatial variability in the growth and development of cotton crops within a field. PA, which comes from the spatial variability distinctive to each field can be measured and managed with site-specific techniques according to the needs of the crop, PA contributes to more effective use of inputs such as fertilizers, pesticides, tillage, and irrigation water. Effective use of inputs will increase crop yield and (or) quality without polluting the environment [7].

To understand fully how PAT is applied, the tools and techniques that create the infrastructure of this modern form of agricultural management need to be well explained. A diagrammatic representation of the PA components is shown in **Figure 1**. It can be said that PA is mainly composed of agricultural data collecting, data processing and analysis, data interpretation and decision making, and variable rate application of inputs [8, 16–18].

The data collecting, which is the first step of PA, entails gathering as much data as possible about crops, soil, fields, terrain, climate, variables, and resource availability by sensors and innovative techniques. Data collecting can be performed using either proximal sensing or remote sensing techniques by special equipment and software such as sensors, GPS technology, controllers, gateways, drones, satellites, and imaging [8].

Data preprocessing and analysis allow for making accurate decisions during variable rate application of inputs by farm machinery because it contributes to a better understanding of crop dynamics, weather, and soil conditions [19]. Information for data interpretation, decision making, and implementation of crop management practices at an appropriate scale and time can be achieved by preprocessing and transforming the raw data acquired through sensing techniques and GIS software [20–22].

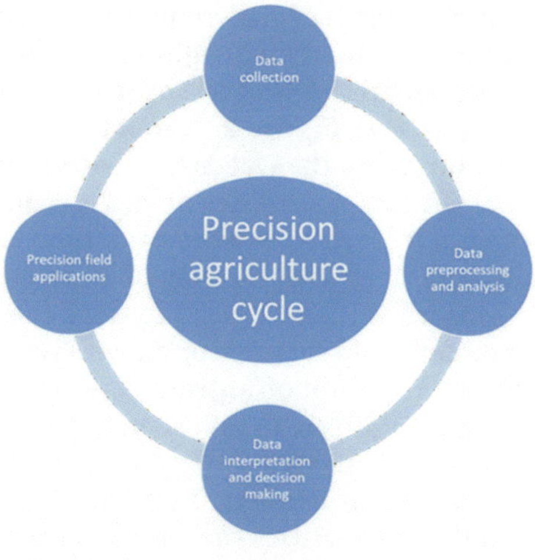

Figure 1.
The main components of precision farming [8, 16, 17].

Data interpretation and decision-making covers choosing the appropriate management tools which give good outcomes in available natural conditions like soil, and environment before using variable-rate devices installed on agricultural equipment. The processes in this phase are considered as very important step in PA [23]. Tantalaki et al. [24] reported that the high volume and complexity of the data caused challenges in successfully implementing PA. They also emphasized that analyzing and interpreting data obtained from ground sensors, unmanned systems, or remote sensing satellites is a significant issue in the successful implementation of PA. The authors state that machine learning techniques, artificial neural networks, support vector machines, decision trees, and random forests, frequently applied for agricultural management purposes, seem promising to cope with agricultural big data, but need to reinvent themselves to meet existing challenges. Also, image segmentation technologies play a significant role in data interpretation and decision making such as plant or weed identification, crop growth stage prediction, crop disease, row detection, and cotton detection in the field [25]. Zhang et al. [26] developed a software used on the smartphone for the real-time detection of cotton diseases and pests in the field. They stated that the developed software could effectively detect the infected area of cotton leaves in the field and provide a technical support for controlling cotton diseases and pests.

Managing field operations can be performed using the information acquired as decision support. Many crop management practices (multiple fertilizer applications, multiple plant growth regulator applications, multiple irrigation potential applications, multiple pesticide applications, etc.) are intensively applied while growing cotton. For this purpose, several agricultural machinery and tools are used for seedbed preparation, sowing, fertilization, pest control weeding, irrigation, and harvesting to reduce labor costs and increase productivity. In PA, these agricultural machinery and tools manage field operations by using technologies such as IoT, AI, remote sensing, and ImP. Recently, numerous robotic systems and variable-rate applicators with human-like capabilities (e.g., precision spraying systems, harvesting robots, shearing robots, grafting machines, weed control systems, transplanting machines, and path planning) have been developed to manage different agricultural activities such as planting, inter-row cultivation, spraying, fertilization, irrigation, and harvesting in crop production [8, 27]. Taylor and Fulton [28] reported that several commercially available sensor-based, variable rate systems exist for efficiently managing inputs to maximize yields or returns. They presented a schematic view of a sensor-based, variable rate application system for liquid products as in **Figure 2**. A robot, which apply a microdose of herbicide systematically targeting the weeds that have been detected, is seen in **Figure 3**. It is stated that the total use of herbicides is reduced by as much as 20 times by using this robot because it detects weeds and then targets weeds by moving independently through the field with the help of a camera, GPS sensor, and a solar drive [29]. Remote sensors are usually used to track the soil conditions and plant health. The data obtained from this sensor allows farmers to selectively use (i.e., precisely apply) the exact amount of nutrients, resources, or pesticides necessary for their fields. Grisso et al. [30], who reviewed the variable rate application devices available on the market, discussed the technologies best fit for a cropping system and production management strategy. They presented an "On-the-go" sensor, which measures soil characteristics such as soil moisture content, texture, electrical conductivity, or soil organic matter before planting and adjusting the

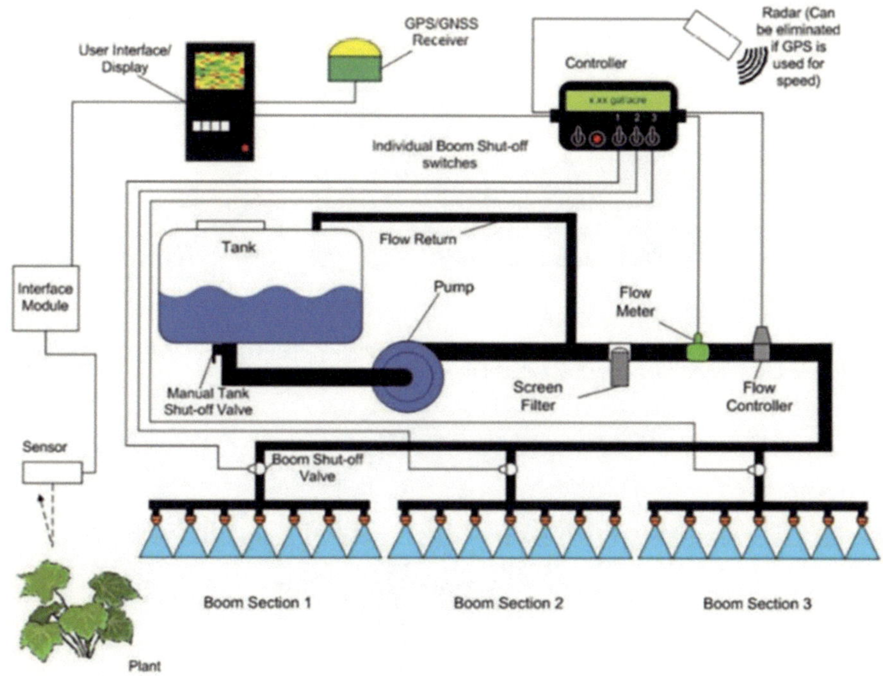

Figure 2.
Schematic of a sensor-based, variable rate application system for liquid products [28].

Figure 3.
A robot, which applies a microdose of herbicide by systematically targeting the weeds detected [29].

seeding rate (plant population) as seen in **Figure 4**. Also, drones and unmanned aerial vehicles systems (UAVs) are used in PA for a variety of tasks such as soil and crop analysis, fertilizer, and pesticide application. Various imaging technologies like hyperspectral, multispectral, and thermal cameras are used in those vehicles,

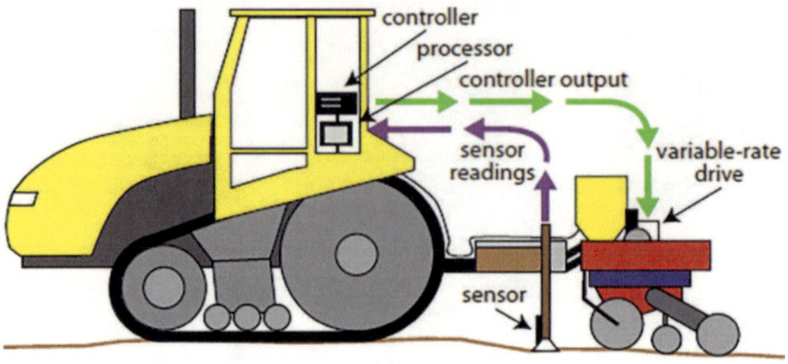

Figure 4.
"On-the-go" sensor (texture, electrical conductivity (EC), or soil organic matter (SOM)) measures soil characteristics before planting and adjusting the seeding rate [30].

Figure 5.
Use of drones and unmanned aerial vehicles systems (UAVs) in precision agriculture [31].

in order to detect and monitor temporal and site-specific changes in plant health and physiology caused by biotic and abiotic stresses [31]. The use of drones and UAVs in precision agriculture is seen in **Figure 5**.

Gemtos et al. [32] stated that a PA system could be divided into three different phases as the acquisition of weather, soil, and crop data; the processing and analysis of the data; and the implementation and adaptation of cultivation practices.

In PA applications, relative observation data and agronomic models were used in order to implement the applications related to tillage and irrigation scheduling, fertilizer management, weed and pest control, soil and crop growth monitoring, and yield estimation. The main aim of PA applications is to apply the targeted rates of fertilizer, seed, and chemicals for soil, crop, and weather conditions by using site-specific knowledge. PAT enable visualization of spatial and temporal variations between fields or within one location and support spatially varying treatments using variable rate application technologies (VRT) installed on farm agricultural field machinery [8, 16, 17, 33]. Variable rate application (VRA) is one of the most important and recent technologies that have been developed recently to accomplish PA. VRA in PA can be map-based or sensor-based. While in map-based VRT, a map of application rates is produced for the field prior to the operation, sensor-based VRT utilize real-time sensors and feedback control to measure the desired properties on-the-go, usually soil properties or crop characteristics, and immediately use this signal to control the variable-rate applicator [33]. Neupane and Guo [34] suggested that the variable rate application of water could reduce water use and improve water use efficiency. Longchamps and Khosla [35] showed that VRA could increase N use efficiency by maintaining productivity and decreasing environmental pollution. Onken and Sunderman [36] determined that the variable rate application of irrigation and fertilization increased cotton yield by 30% when compared to whole-surface application. The usage of PA or site-specific management applications in cotton production fields reduces this variation in yield by recognizing field spatial variability and optimizing variable input use within fields [37]. Huang et al. [38], who reviewed the remote sensing technologies available on the market for weed management, presented in detail information on the development and application of UAVs-based low-altitude remote sensing technology for precision weed management. Lamm et al. [39] developed a real-time robotic weed control system and tested it in commercial cotton fields. The researchers stated that this precision weed control system, which consisted of a real-time machine vision system, a controlled illumination chamber, and a precision chemical applicator, was capable of distinguishing grass-like weeds from cotton plants and applying a chemical spray only to targeted weeds. Allmendinger et al. [40] summarized different commercial technologies and prototypes for precision patch spraying and spot spraying. The authors presented an overview of sensors, applications, and implementation options that should be possible to be controlled via ISOBUS-Connection, as seen in **Figure 6**. Also, PA facilitates other management decisions making, such as site-specific deep tillage to remove soil compaction, and the equipment guidance where farm equipment follows the same paths for various field operations [41].

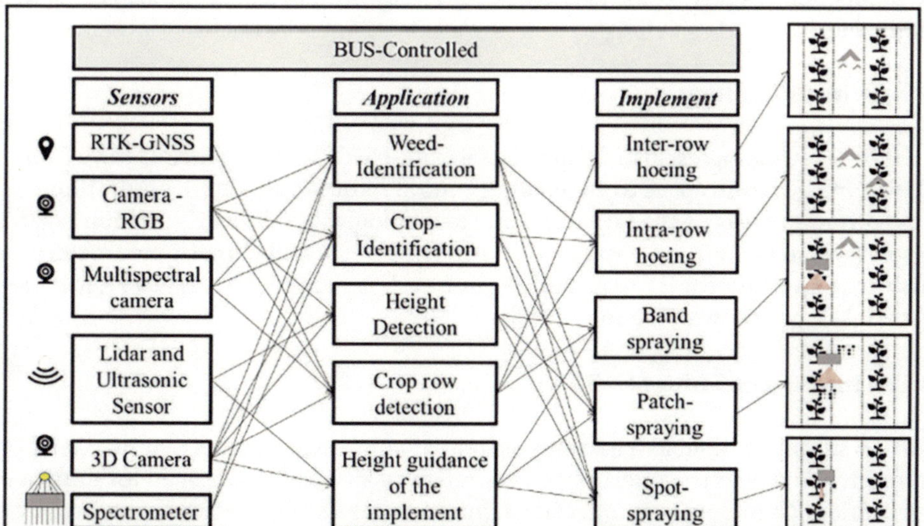

Figure 6.
Overview of sensors, application, and implementation options that should be possible to be controlled via ISOBUS-connection [40].

3. Apparatus and instruments in precision agriculture

PA requires special equipment and software to collect and analyze all the information. It uses a range of technologies or diagnostic tools such as GPS, GIS, yield monitors, near-infrared reflectance sensing, remote sensing, IoT sensors, drones, computer vision, LIDAR, big data processing, and artificial intelligence in order to collect and analyze the in-field spatial variability data, and after then, make site-specific management decisions for soils and crops [8, 42, 43].

3.1 Global positioning system (GPS)

GPS, one of the most important parts of PA, enables precise recording of coordinates (latitude and longitude) to accurately map and pinpoint the location of the device or sensor as the data is collected and then aggregates all of these locations and data to create a visual geographic map. GPS is widely used on drones, sensors, tractors, and other farm machinery. GPS enables machinery to operate autonomously and aid in soil testing, tractor driving with a parallel steering system, and VRA for precise seed and fertilizer application.

3.2 Geographic information system (GIS)

GIS, one of the key tools for PA, is a digital mapping system that manages the spatial information collected on the ground. It is also defined as a software that collects, analyzes, and displays spatial data on maps. GIS can help farmers to use various types of data and analytics to improve farmer's decision making and planning in many ways. For example, GIS can help the farmer to map and monitor the soil characteristics and variability in their fields, such as texture, pH, organic matter, moisture, and fertility. After that, by integrating soil data with other information, such as weather, crop type, and yield, farmers can create soil management zones and apply variable rate inputs, such as fertilizers, pesticides, and irrigation, to each zone according to its specific needs.

3.3 Satellite technology

Satellite technologies, also defined as satellite-based remote sensing, have been utilizing images from space to analyze vegetation indices and identify variations across different areas of the farm. Satellite technologies provide information regarding soil type and condition, crop health, and hydrologic and climatic parameters, which are important for PA (e.g., soil organic carbon, soil moisture, NDVI, leaf area index (LAI), groundwater, and rainfall).

3.4 The Internet of Things (IoT)

IoT is simply defined as a network system of connected devices that may send and receive data over the internet and complete tasks without the need for human intervention. Its main aim is to create an internet-based huge network platform by combining several sensor technologies and networks, in order to understand how information is shared among items all over the world.

3.5 Auto-guidance system

Auto-guidance system is a computer-controlled system that helps farmers guide their tractors or other agricultural equipment along the desired path. The system uses GPS technology to track the equipment's location and automatically steer it in the desired direction. There are many different types of auto-guidance systems (GPS-based auto-guidance systems, laser-based auto-guidance systems, camera-based auto-guidance systems) available on the market, each with its advantages and disadvantages.

3.6 Precision agriculture-based soil sampling

Soil sampling helps to evaluate the physical and nutrient status of soil in a field to base crop management decisions such as planting, cultivation, fertilization, and irrigation for optimal crop production. Precision soil sampling results in higher fertilizer use efficiency, reduced nutrient loss, and protection of the surrounding natural resources [44, 45].

3.7 Variable-rate technology (VRT)

VRT is a technology which adjusts and applies precise and measured the quantities of the inputs (e.g., fertilizers, pesticides, herbicides, seeds, and water) in agricultural production by considering data that is collected by sensors, maps, and GPS in a given landscape. VRT consists of the machines and systems for applying a desired rate of crop production materials at a specific time and a specific location; a system of sensors, controllers, and agricultural machinery used to perform variable-rate applications of crop production inputs.

3.8 Crop and soil sensors/remote sensing

Crop and soil sensors/remote sensing technologies are mainly used to measure essential soil and crop properties on-the-go in the field. The measurement results obtained from these sensors are used either to control variable rate application equipment in real-time or in conjunction with a global positioning system (GPS) to generate field maps. There are numerous sensor types (e.g., location sensors, optical sensors, multispectral sensors, thermal infrared sensors, laser sensors, electro-chemical sensors, mechanical sensors, dielectric soil moisture sensors, and air flow sensors) on the market to measure crop, soil, and microclimate parameters. Kim and Lee [46], who reviewed the sensors widely used in the market in order to monitor the plant, the soil, and the environmental conditions that directly affect plant growth, presented the electrochemical sensors for monitoring plant health in PA as in **Figure 7**.

3.9 Image-based sensing and unmanned aerial vehicles

Unmanned aerial vehicles (UAV), and unmanned ground vehicles (UGV) are utilized for plant health monitoring, pest control, livestock management, aerial survey, and soil analysis in PA. High-resolution cameras (RGB, multispectral, etc.) are used for the purpose of capturing the images for further investigation [47].

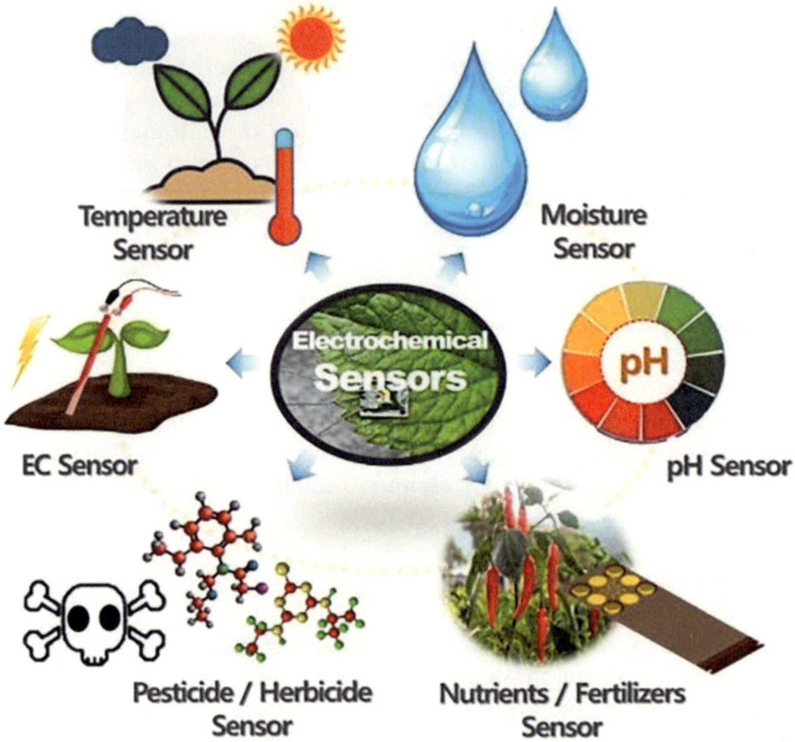

Figure 7.
The electrochemical sensors for monitoring plant health in precision agriculture [46].

3.10 Crop yield monitoring and mapping

Crop yield monitoring and mapping can be defined as the process of collecting georeferenced data on crop yield and characteristics while the crop is being harvested. It provides the ability to not just estimate yield, but to identify the location in the field where yield is produced. The information obtained during crop yield monitoring and mapping is valuable for a multitude of management purposes, including estimating the amount of nutrients removed by the harvested crop, estimating profitability, developing management zones, and analyzing the impacts of treatments used in on-farm studies. In general, yield monitors provide a realistic estimate of the "relative" yield differences within a field. Yield maps are very useful in providing a visual image which shows the variability of yield across a field [48–50]. Various methods, using a range of sensors, have been developed for mapping crop yields. Vellidis et al. [49], who evaluate the strengths and weaknesses of commercially available crop yield monitoring and mapping systems, recommended that all potential users should carefully research prospective cotton yield monitoring systems for the following attributes before purchasing: quality of the product, "user-friendliness," ease of installation, GPS requirements, availability and responsiveness of technical support, skill level required of the picker operator, and time available for downloading data files.

4. Main issues affecting adoption of precision agriculture in cotton production

Although PAT improve the efficiency of agricultural practices by reducing overuse of inputs (seed, fertilizer, pesticides, etc.), thus saving money on input costs, it is seen that adoption of PAT remains relatively low in cotton production. Many studies [e.g., [11, 51–57]] have been carried out to analyze the factors affecting the adoption of PAT specifically among cotton producers in recent years since it is stated that PAT are generally more profitable with high-value crops, such as cotton [58]. These studies showed that many factors including socio-demographic, economic, institutional infrastructure, etc., affect the adoption of PAT. The most important of these factors are expressed as the lack of digital infrastructure like internet and electricity, education level and purchasing power of farmers, and societal barriers. Pandey et al. [8] presented some common issues affecting the adaption of PA as data management, hardware cost, lack of information, interoperability, connectivity, and environmental variation (**Figure 8**).

Yield monitors are considered as one of the most important technologies in PA [59]. Also, grid and zone soil sampling and the use of soil survey maps are considered as a significant entry technology into PA [60]. However, the adoption sequence of PAT in cotton production is different from that in grain production because of the lack of reliable yield monitoring technologies for cotton [54, 61]. For example, reliable yield monitors for cotton were not available until 2000, while monitors for grains and oilseeds have been on the market since the early 1990s [62]. Also, in the USDA (US Department of Agriculture) ERS (Economic Research Service), it is stated that yield monitors were used on 72% of the corn area planted in 2010 and 33% of the corn area was mapped using yield monitor-GPS systems, in contrast, it was used on only 4.7% of the cotton area planted in 2007 and only 2.8% of the cotton area planted was mapped with yield monitors [63]. Mooney et al. [64] observed the adoption rate of cotton yield monitors by farms in 2009 as 4%. Larson et al. and Boyer et al. [61, 65] observed the use of yield monitors with GPS in cotton production has risen from

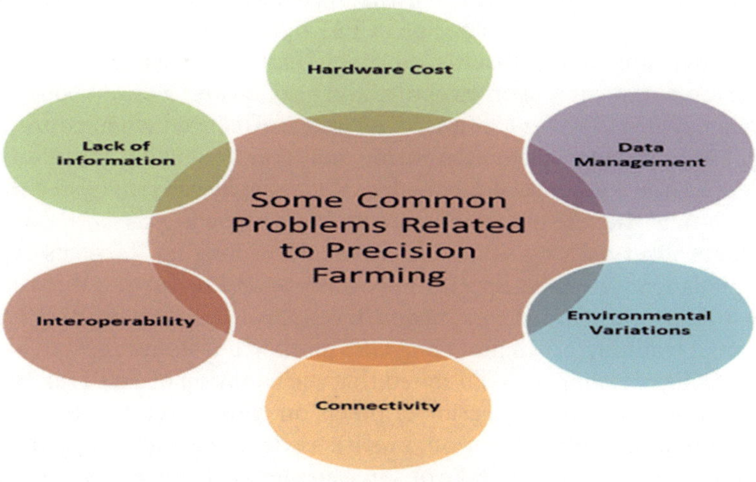

Figure 8.
Some common issues affecting the adaption of precision agriculture [8].

2.8% in 2001 to 19% in 2013. This situation significantly affected the adoption and frequency of PAT in cotton production.

Reuters [66] stated that the most important factor influencing the adoption of cotton yield monitors might be the introduction of on-board module builders on cotton harvesters that are paired with yield monitoring technology. Martin and Varco [67] reported that farmers might find value in combining the two technologies because of reduced equipment and labor expenses associated with the elimination of boll buggies and module builders in the harvest equipment complement. Roberts et al. [68] stated that cotton producers' first experience in precision agriculture started with precision soil sampling, not yield monitoring. Walton et al. [69], who analyzed cotton farmer decisions regarding the adoption and abandonment of precision soil sampling as a function of farm and farmer characteristics, observed that farmers with high levels of education, large cotton production areas, and digital skills had an optimistic opinion about future of PA and were more likely to adopt precision soil sampling for cotton production. Lambert et al. [54] analyzed the adaptation of precision soil sampling by cotton producers in thirteen southern states in 2013. They stated that farming experience, farm size, land ownership, variable rate fertilizer, management plans, and the use of soil electrical conductivity devices had significant effects on the adaptation of precision soil sampling. Shafi et al. [70] stated that PA has been used for the last few decades to enhance crops' yield with reduced costs and human effort, although the adoption of these novel techniques by farmers is still very limited owing to the reasons or challenges such as hardware cost, weather variations, data management, literacy rate, connectivity, and interoperability. Esposito et al. [71] reviewed the potential and practical use of the most advanced sensors available in the market for precision weed control. They emphasized that nowadays, especially, PA has rapidly advanced in integrated weed management because of technological innovations in the areas of sensors, computer hardware, nanotechnology, unmanned vehicle systems, and robots that may allow for specific identification of weeds that are present in the field. Lambert et al. [72], who evaluate the factors influencing the timing of grid soil sampling, yield monitoring, and remote sensing adoption by cotton producers using multivariate censored regression, stated that understanding the factors influencing the early adoption of PAT by cotton farmers is important for anticipating technology diffusion over time. They suggested that different factors such as land ownership, farm structure and size, farmer age and education, the purchasing behavior of farmers, and farm location influenced when cotton farmers adopted grid soil sampling, yield monitoring, and remote sensing adoption after these technologies became commercially available. Khanal et al. [73] suggested that special attention is given to the role of farmer expectations, following the adoption of PAT. The researchers observed a significant positive role of meeting "farmer's expectation" about GPS guidance systems in application decisions and its further diffusion within a cotton farm. Also, they stated that income level, farm size, and farming occupation were other important factors in modeling GPS guidance system adoption and application. Takács-György et al. [74], who stated that the adoption of PAT is slow across the world, emphasized that the application of precision crop production is not easy to understand, it requires much attention, precise work, and a wide range of information. They stated that the slow uptake of some elements of the technology could be partly explained by the problematic questions of shifting such as the need for expertise and precision, requiring the documentation and tracking of the procedures, and extra investment. The authors suggested that all kinds of cooperation and strategic collaborations among the farmers, extension services, and providers are important in the adoption of new technology, such as the forms of joint machine use (e.g., machinery rings) because the individuals make their decisions on the adoption of new technologies on the basis of information coming through these channels. This means that relevant industry players and suppliers affected in the application and marketing of the technology are dominant with regard to the application and adoption. Nair et al. [75] determined that farm size, extension activities, percentage of land owned by a farmer, and the age-education had significant impacts on the choice of the VDTs and VRT. They stated that in particular, younger and more educated farmers were more likely to adopt VDRs and VRT.

A recent study by Paudel et al. [76], who analyzed the duration of PAT adoption that addresses heterogeneity and event dependence between PAT adoptions among the US cotton farmers. Their results indicated that farmers with large farms, a higher share of total cultivated farmland, a higher percentage of income from farming, and farmers using computers for farm management were more likely to adopt PAT. Further, the researchers observed that cotton producers who think that PAT would be valuable in the future and those receiving farming information from university publications were more likely to adopt PAT soon after the technologies become available. In order to increase the adoption of PAT, they recommend the following remarks:

1. Political and legal support could be provided for the dissemination of precision agriculture.

2. University extension programs, private extension services, or international extension agents could provide valuable educational and application assistance to producers to become more familiar with the usage of the PAT.

3. Subsidized capital or financial support could be provided to facilitate the adoption process of farmers because the equipment used in PAT is expensive.

4. Proper documents of environmental or agronomic benefits from precision agriculture might contribute farmers to adopt the PAT.

Overall, we can summarize the main deterrents to the widespread adoption of PA as follows;

1. Considerable investments in time and effort are usually required to learn how to use new technologies in the case of adopting PA.

2. The lack of demonstrated evidence for the economic advantages of adopting PA.

3. Uncertainty in returns and high fixed cost prevents adoption of precision agriculture applications.

4. Farmers' lack of awareness of the existing precision agriculture technologies in the market.

5. Difficulty in understanding the technologies and interpretation of the data.

6. Farm size, exposure to extension activities, and the age-education have a significant impact on the adoption of PAT.

5. Conclusions

Precision agriculture (PA), which aims to optimize crop yields and reduce waste, while minimizing the impact on the environment by applying crop input and agronomic practices according to the spatial and temporal variability in field conditions and crop requirements, is a new or latest farming practice. It can also be defined as an agricultural application using modern technologies such as GPS, GIS, yield monitors, near-infrared reflectance sensing, remote sensing, IoT sensors, drones, computer vision, LIDAR, big data processing, and artificial intelligence. PA is still in its infancy and its adoption varies greatly although it is the agricultural system of the future and has tremendous benefits. This chapter has given an extensive literature survey on PA and its practices in cotton production along with several considerations and challenges. Also, the information about the main apparatus and instruments used in PA are given in the chapter. Furthermore, the main factors affecting the adoption of PA in cotton production are highlighted. According to the results of the articles reviewed within the scope of this chapter, the adoption sequence of PAT in cotton production is different from that in grain production because of the lack of reliable yield monitoring technologies for cotton. The studies showed that many factors including socio-demographic, economic, institutional infrastructure, etc., affected the adoption of PAT. In the studies, some common issues affecting the adaption of PA are presented as data management, hardware cost, lack of information, interoperability, connectivity, environmental variation, uncertainty in returns from adoption, high fixed cost, farmers' lack of awareness of the existing PAT in the market, farm size, exposure to extension activities, and the age-education.

Author details

Songül Gürsoy
Department of Agricultural Machinery and Technologies Engineering, Dicle University, Diyarbakır, Turkey

*Address all correspondence to: songulgursoy@hotmail.com

References

[1] Khan MA, Wahid A, Ahmad M, Tahir MT, Ahmed M, Ahmad S, et al. World cotton production and consumption: An overview. In: Ahmad S, Hasanuzzaman M, editors. Cotton Production and Uses. Singapore: Springer; 2020. pp. 1-7

[2] OECD/FAO. Agricultural Outlook 2021-2030. Paris, France: OECD Publishing; 2021. p. 337. DOI: 10.1787/19428846-en

[3] Galanopoulou-Sendouca S. Towards a more competitive and sustainable cotton production. In: Lekakis JN, editor. Freer Trade, Sustainability, and the Primary Production Sector in the Southern EU: Unraveling the Evidence from Greece, Economy and Environment. Vol. 16. Netherlands, Dordrecht: Springer; 1998. pp. 112-130

[4] Robert PC. Site-specific management for the twenty-first century. HortTechnology. 2000;**10**(3):444-447

[5] Bouma J, Stoorvogel J, van Alphen BJ, Booltink HWG. Pedology, precision agriculture, and the changing paradigm of agricultural research. Soil Science Society of America Journal. 1999;**63**:1763-1768

[6] Kırkaya A. Smart farming- precision agriculture technologies and practices. Journal of Scientific Perspectives. 2020;**4**(2):123-136

[7] Davis G, Casady W, Massey R. Precision Agriculture: An Introduction [Internet]. 1998. Available from: https://extension.missouri.edu/publications/wq450 [Accessed: July 08, 2023]

[8] Pandey H, Singh D, Das R, Pandey D. Precision farming and its application. In: Choudhury A, Biswas A, Singh TP, Ghosh SK, editors. Smart Agriculture Automation Using Advanced Technologies, Transactions on Computer Systems and Networks. Singapore: Springer; 2021. pp. 17-33

[9] Heermann DF, Hoeting J, Thompson SE, Duke HR, Westfall DG, Buchleiter GW, et al. Interdisciplinary irrigated precision farming research. Precision Agriculture. 2002;**3**(1):47-61

[10] Koch B, Khosla R, Frasier WM, Westfall DG, Inman D. Economic feasibility of variable-rate nitrogen application utilizing site-specific management zones. Agronomy Journal. 2004;**96**:1572-1580

[11] Roberts RK, English BC, Larson JA, Cochran RL, Goodman WR, Larkin SL, et al. Adoption of site-specific information and variable-rate technologies in cotton precision farming. Journal of Agricultural and Applied Economics. 2004;**36**(1):143-158. DOI: 10.1017/S107407080002191X

[12] Paxton KW, Mishra AK, Chintawar S, Roberts RK, Larson JA, English BC, et al. Intensity of precision agriculture technology adoption by cotton producers. Agricultural and Resource Economics Review. 2011;**40**(1):133-144

[13] D'Antoni JM, Mishra A, Joo H. Farmers' perception of precision technology: The case of autosteer adoption by cotton farmers. Computers and Electronics in Agriculture. 2012;**87**:121-128

[14] Karaş E. In: Abdurakhmonov IY, editor. Sustainable and Effective Management Strategies in Cotton Cultivation. London: IntechOpen; 2022. DOI: 10.5772/intechopen.104104

[15] Razaq M, Mensah R, Athar HR. Insect pest management in cotton. In: Jabran K, Chauhan BS, editors. Cotton Production. 1st ed. New York: Wiley; 2020. pp. 85-107

[16] Blackmore BS, Wheeler PN, Morris RM, Morris J, Jones RJA. The role of precision farming in sustainable agriculture: A European perspective. In: Robert PC, Rust RH, Larson WE, editors. Site-Specific Management for Agricultural Systems; Madison: Wley; 1995. pp. 77-793

[17] Blackmore S, Griepentrog HW, Pedersen SM, Fountas S. Precision farming in Europe. In: Srinivasan A, editor. Handbook of Precision Agriculture: Principles and Applications. Newyork, London, Oxford: Haworth Press; 2006. pp. 567-613

[18] National Research Council. Precision Agriculture in the 21st Century: Geospatial and Information Technologies in Crop Management. Washington, DC: National Academy Press; 1997. pp. 1-168

[19] Liakos K, Busato P, Moshou D, Pearson S, Bochtis D. Machine learning in agriculture: A review. Sensors. 2018;**18**:2674. DOI: 10.3390/s18082674

[20] Nandyala C, Kim HK. Big and meta data management for U-agriculture mobile services. International Journal of Software Engineering and Its Applications. 2016;**10**:257-270. DOI: 10.14257/ijseia.2016.10.2.21

[21] Wolfert S, Ge L, Verdouw C, Bogaardt MJ. Big data in smart farming – A review. Agricultural Systems. 2017;**153**:69-80. DOI: 10.1016/j.agsy.2017.01.023

[22] Cravero A, Sepúlveda S. Use and adaptations of machine learning in big data—Applications in real cases in agriculture. Electronics. 2021;**10**(5):552

[23] Akhter R, Sofi SA. Precision agriculture using IoT data analytics and machine learning. Journal of King Saud University - Computer and Information Sciences. 2022;**34**(8-Part B):5602-5618

[24] Tantalaki N, Souravlas S, Manos R. Data-driven decision making in precision agriculture: The rise of big data in agricultural systems. Journal of Agricultural and Food Information. 2019;**20**(4):344-380

[25] Li Y, Cao Z, Lu H, Xiao Y, Zhu Y, Cremers AB. In-field cotton detection via region-based semantic image segmentation. Computers and Electronics in Agriculture. 2016;**127**:475-486

[26] Zhang Y, Ma B, Hu Y, Li C, Li Y. Accurate cotton diseases and pests detection in complex background based on an improved YOLOX model. Computers and Electronics in Agriculture. 2022;**203**:107484

[27] Fue KG, Porter WM, Barnes EM, Rains GC. An extensive review of mobile agricultural robotics for field operations: Focus on cotton harvesting. AgriEngineering. 2020;**2**(1):150-174

[28] Taylor R, Fulton J. Sensor–based variable rate application for cotton [Internet]. 2021. Available from: https://cottoncultivated.cottoninc.com/wp-content/uploads/2015/05/Sensor-Based-Variable-Rate-Application.pdf [Accessed: September 21, 2023]

[29] Zimmer D, Plaščak I, Barač Ž, Jurišić M, Radoča D. Application of robots and robotic systems in agriculture. Tehnıčkı Glasnık. 2021;**15**(3):435-442

[30] Grisso RB, Alley M, Thomason W, Holshouser D, Roberson GT. Precision farming tools: Variable-rate application [Internet]. 2011. Available from:

https://vtechworks.lib.vt.edu/bitstream/handle/10919/47448/442-505_PDF.pdf [Accessed: September 21, 2023]

[31] Raj R, Kar S, Nandan R, Jagarlapudi A. Precision agriculture and unmanned aerial vehicles (UAVs). In: Avtar R, Watanabe T, editors. Unmanned Aerial Vehicle: Applications in Agriculture and Environment. Cham: Springer; 2020. pp. 7-23. DOI: 10.1007/978-3-030-27157-2_2

[32] Gemtos TA, Ath M, Toulios L, Pateras D, Zerva G. Precision farming applications in cotton fields of Greece. In: Proceedings of CIGR International Conference; October 11-14, 2004; China: CIGR, CSAM and CSAE; 2004. pp. 10-14

[33] Igor I. Variable rate application in precision agriculture [Internet]. 2018. Available from: https://medium.com/remote-sensing-in-agriculture/variable-rate-application-in-precision-agriculture-70a8b2be871d [Accessed: September 21, 2023]

[34] Neupane J, Guo W. Agronomic basis and strategies for precision water management: A review. Agronomy. 2019;**9**(2):87

[35] Longchamps L, Khosla R. Improving N use efficiency by integrating soil and crop properties for variable rate N management. In: Stafford JV, editor. Precision Agriculture. 1st ed. Washington: Wageningen Academic Publishers; 2015. pp. 249-256

[36] Onken AB, Sunderman HD. Applied and residual nitrate-nitrogen effects on irrigated grain Sorghum yield. Soil Science Society American Proceedings. 1972;**36**:94-96

[37] Snyder CJ. An economic analysis of variable-rate nitrogen management using precision farming methods [thesis]. Kansas: Kansas State University; 1996

[38] Huang Y, Reddy KN, Fletcher RS, Pennington D. UAV low-altitude remote sensing for precision weed management. Weed Technology. 2018;**32**(1):2-6

[39] Lamm RD, Slaughter DC, Giles DK. Precision weed control system for cotton. Transactions of the ASAE. 2002;**45**(1):231-238

[40] Allmendinger A, Spaeth M, Saile M, Peteinatos GG, Gerhards R. Precision chemical weed management strategies: A review and a design of a new CNN-based modular spot sprayer. Agronomy. 2022;**12**(7):1620

[41] Lowenberg-DeBoer J, Erickson B. Setting the record straight on precision agriculture adoption. Agronomy Journal. 2019;**111**(4):1552-1569

[42] Sahu B, Chatterjee S, Mukherjee S, Sharma C. Tools of precision agriculture: A review. International Journal of Chemical Studies. 2019;7(6):2692-2696

[43] Meena BR, Dudwal BL. Precision farming; their tools and techniques. Just Agriculture. 2021;**2**(1):1-11

[44] Ferguson RB, Hergert GW. Soil sampling for precision agriculture [Internet]. 2009. Available from: https://extensionpublications.unl.edu/assets/pdf/ec154.pdf [Accessed: July 08, 2023]

[45] Harper DC. Soil test information in cotton production: Adoption, use, and value in potassium management [thesis]. Tennessee: The University of Tennessee; 2011

[46] Kim MY, Lee KH. Electrochemical sensors for sustainable precision agriculture—A review. Frontiers in Chemistry. 2022;**10**:848320

[47] Maes WH, Steppe K. Perspectives for remote sensing with unmanned aerial

vehicles in precision agriculture. Trends in Plant Science. 2019;**24**(2):152-164

[48] Wilkerson JB, Moody FH, Hart WE. Implementation and field evaluation of a cotton yield monitor. Applied Engineering in Agriculture. 2002;**18**(2):153-159

[49] Vellidis G, Perry C, Wells T. Cotton yield monitors the entrance exam and final exam for precision agriculture [Internet]. 2017. Available from: https://www.cottoninc.com/wp-content/uploads/2017/03/cottonyieldmonitors.pdf [Accessed: July 28, 2023]

[50] Fulton J, Hawkins E, Taylor R, Franzen A. Yield monitoring and mapping. In: Shannon DK, Clay DE, Kitchen NR, editors. Precision Agriculture Basics; Madison: Wiley; 2018. pp. 63-78

[51] Watcharaanantapong P, Roberts RK, Lambert DM, Larson JA, Velandia M, English BC, et al. Timing of precision agriculture technology adoption in US cotton production. Precision Agriculture. 2014;**15**:427-446

[52] Tenkorang F, Lowenberg-DeBoer J. On-farm profitability of remote sensing in agriculture. Journal of Terrestrial Observation. 2008;**1**(1):50-59

[53] Daberkow SG, McBride WD. Farm and operator characteristics affecting the awareness and adoption of precision agriculture technologies in the US. Precision Agriculture. 2003;**4**:163-177

[54] Lambert DM, Paudel KP, Larson JA. Bundled adoption of precision agriculture technologies by cotton producers. Journal of Agricultural and Resource Economics. 2015;**40**(2):325-345

[55] Paxton KW, Mishra AK, Chintawar S, Roberts RK, Larson JA, English BC, et al. Intensity of precision agriculture technology adoption by cotton producers. Agricultural and Resource Economics Review. 2011;**40**(01):133-144

[56] Larson JA, Roberts RK, English BC, Larkin SL, Marra MC, Martin SW, et al. Factors affecting farmer adoption of remotely sensed ımagery for precision management in cotton production. Precision Agriculture. 2008;**9**(4):195-208

[57] Walton JC, Larson JA, Roberts RK, Lambert DM, English BC, Larkin SL, et al. Factors influencing farmer adoption of portable computers for site-specific management: A case study for cotton production. Journal of Agricultural and Applied Economics. 2010;**42**(2):193-209

[58] Swinton SM, Lowenberg-DeBoer J. Evaluating the profitability of site-specific farming. Journal of Production Agriculture. 1998;**11**(4):439-446

[59] Lowenberg de Boer J. Adoption of GPS based guidance systems in agriculture [Internet]. 1999. Available from: http://dynamo.ecn.purdue.edu/~biehl/SiteFarmin/publications.html [Accessed: October 15, 2023]

[60] Schimmelpfennig D, Ebel R. On the doorstep of the information age: Recent adoption of precision agriculture. Economic Information Bulletin, U.S. Department of Agriculture, Economic Research Service, Washington, DC, 2011. Available from: http://purl.fdlp.gov/GPO/gpo15242

[61] Larson JA, Roberts RK, English BC, Cochran RL, Wilson BS. A computer decision aid for the cotton yield monitor investment decision. Computers and Electronics in Agriculture. 2005;**48**(3):216-234

[62] Perry CD, Vellidis G, Wells N, Kvien C. Simultaneous evaluation of

multiple commercial yield monitors in Georgia. In: Proceedings of Beltwide Cotton Conference; January 9-13, 2001; Anaheim, CA. Memphis, TN: National Cotton Council; 2001. pp. 328-338

[63] USDA (US Department of Agriculture) ERS (Economic Research Service). Agricultural resource management survey farm financial and crop production practice [Internet]. 2010. Available from: http://www.ers.usda.gov/data-products/arms-farm-financial-and-crop-production-practices.aspx [Accessed: June 12, 2023]

[64] Mooney DF, Roberts RK, English BC, Lambert DM, Larson JA, Velandia M, Larkin SL, Marra MC, Martin SW, Mishra A, Paxton KW, Rejesus R, Segarra E, Wang C, Reeves JM. Precision farming by cotton producers in twelve southern states: Results from the 2009 southern cotton precision farming survey [Internet]. 2010. Available from: https://ageconsearch.umn.edu/record/91333/files/2009%20Cotton%20Precision%20Farming%20Research%20Report.pdf [Accessed: August 12, 2023]

[65] Boyer CN, English BC, Roberts R, Larson J, Lambert DM, Velandia MM, Zhou V, Larkin SL, Marra MC, Rejesus RM, Falconer LL, Martin SW, Mishra AK, Paudel KP, Wang C, Johnson J, Segarra E, Reeves JM. Results from a Cotton Precision Farming Survey Across Fourteen Southern States [Internet]. 2014. Available from: https://ncc.confex.com/ncc/20 1 4/webprogram/Paper 1 4957.html [Accessed: August 12, 2023]

[66] Reuters. Case IH Module Express 625 Streamlines Cotton Harvest [Internet]. 2008. Available from: http://www.reuters.com/article/2008/01/08/idUS130336+08-Jan-2008+MW20080108 [Accessed: August 12, 2023]

[67] Martin SW, Valco TD. Economic comparison of on-board module builder harvest methods. In: Proceedings of 2008 Beltwide Cotton Conferences; August 12-14, 2008; Memphis, TN: National Cotton Council; 2008. pp. 802-804

[68] Roberts RK, English BC, Larson JA, Cochran RL, Goodman B, Larkin S, Marra MC, Martin S, Reeves J, Shurley D. Precision farming by cotton producers in six southeastern states: Results from the 2001 southern precision farming survey [Internet]. 2002. Available from: https://citeseerx.ist.psu.edu/document?repid=rep1&type=pdf&doi=a410ebac85fb43688bf4e995820aff65c2c8611d [Accessed: August 12, 2023]

[69] Walton JC, Lambert DM, Roberts RK, Larson JA, English BC, Larkin SL, et al. Adoption and abandonment of precision soil sampling in cotton production. Journal of Agricultural and Resource Economics. 2008;**33**(3):428-448

[70] Shafi U, Mumtaz R, García-Nieto J, Hassan SA, Zaidi SAR, Iqbal N. Precision agriculture techniques and practices: From considerations to applications. Sensors. 2019;**19**(17):3796. DOI: 10.3390/s19173796

[71] Esposito M, Crimaldi M, Cirillo V, Sarghini F, Maggio A. Drone and sensor technology for sustainable weed management: A review. Chemical and Biological Technologies in Agriculture. 2021;**8**:18

[72] Lambert DM, English BC, Harper DC, Larkin SL, Larson JA, Mooney DF, et al. Adoption and frequency of precision soil testing in cotton production. Journal of Agricultural and Resource Economics. 2014;**39**(1):106-123

[73] Khanal AR, Mishra AK, Lambert DM, Paudel KP. Modeling post adoption decision in precision

agriculture: A Bayesian approach. Computers and Electronics in Agriculture. 2019;**162**:466-474

[74] Takács-György K, Lencsés E, Takács I. Economic benefits of precision weed control and why its uptake is so slow. Studies in Agricultural Economics. 2013;**115**:40-46

[75] Nair S, Wang C, Segarra E, Belasco E, Larson J, Velandia M, et al. Adoption of precision agriculture for cotton in the southern United States. Journal of Agribusiness. 2011;**29**(2):221-243

[76] Paudel KP, Mishra AK, Pandit M, Segarra E. Event dependence and heterogeneity in the adoption of precision farming technologies: A case of US cotton production. Computers and Electronics in Agriculture. 2021;**181**:105979

Chapter 7

The Strategic Importance of Cotton Production for the World and Türkiye

Songül Akın

Abstract

Cotton can be grown in a wide variety of geographical regions worldwide. It is a valuable crop as it contributes positively to the economy of the regions and countries with its wide range of use, added value and employment opportunities. Beyond its being just an agricultural commodity, cotton has strategic importance for Türkiye where there have been ideal regions for cultivation. Growing interest in natural fiber and rising living standards increase the demand for cotton. In this study, in addition to the general information on cotton crops and their cultivation, explanations and interpretations were made on production and trade particularly in Türkiye and trade generally in the world for this crop based on the recent cotton production and trade data.

Keywords: cotton production, cotton trade, strategic product, employment, Türkiye

1. Introduction

Agriculture continues to be the basic sector to meet the primary needs of human beings. It is also of very importance regarding the economic development of the countries and raising the general welfare level. This sector has been considered the cornerstone of the economy for many years, yet the developments in the industrial sector in recent years have begun to be interpreted as the shift of importance from agriculture to industry. This interpretation can be acceptable, especially for technology-producing developed countries, but it is open to discussion for developing countries like Türkiye.

Considering its high agricultural production potential, and a significant share of industry based on agriculture, the importance of this sector becomes more obvious for Türkiye. Human nutrition, food security, resource transfer to industry and income distribution features of agriculture come to the fore to consider it as a strategic sector [1].

The branches of industry that process agricultural raw materials in Türkiye consist of two main groups, the food and the textile industries, which are included in the manufacturing industry. Sugar, vegetable oil, flour, canned food, milk, meat and beverage industries are the sub-categories of the food industry as cotton thread, wool yarn, cotton and woolen weaving are the sub-branches of the textile industry [2].

The increasing world population has affected the food sector as well as many other sectors. The ever-increasing needs due to the change in living standards have also led to an increase in interest in the textile sector. The fact that cotton is the basic raw material of the textile industry in Türkiye as in the world, increases the importance of this crop day by day.

Parallel to an upward shift of public awareness on artificial fiber pollution and its damage to the environment due to their long lifecycle in nature; especially in recent years, the rise in demand for cotton, the largest natural fiber source, has caused the market share of cotton to increase again.

Cotton is one of the main agricultural crops widely grown and traded in the world [3]. Plant height varies between 80 and 120 cm and boll branches are separate, it is a tap-rooted plant with a long flowering and harvesting period It is an annual crop with an effective root zone depth of 90 cm and can be grown under warm climate and irrigated conditions [4].

Mainly four cotton species (*Gossypium hirsutum*, *Gossypium barbadense*, *Gossypium arboretum* and *Gossypium herbaceum*) are grown worldwide. *Gossypium hirsutum*, also called upland cotton, makes up of more than 90% of the global cotton production and 99% of the Turkish cotton production. Upland cotton (*Gossypium hirsutum* L.) have a high yield potential, a medium-long vegetation period with a ginning percentage of about 39% [5].

Due to the deterministic effect of climate cotton is grown generally in the Southeastern Anatolia, Aegean and Mediterranean regions in Türkiye. The lint percentage of seed cotton (lint plus seed) is about 35–40% and the rest is the seed. The oil content of cottonseeds varies between 19 and 28%. Although cotton is generally grown for its fiber, cottonseed oil ranks fourth after soybean, rapeseed and sunflower in terms of world oil production. It also ranks second after soy as a plant protein source [6].

The price elasticity of demand for cotton is lower than that of other industrial crops due to its wide range and frequent use.

With the advancing technology, cotton fiber is used in textile industry, linter is used in paper, furniture and cellulose industry, seed is used in seed oil industry, and pulp is used in feed industry [7]. For that reason, it not only has a high commercial value but also has a high employment potential globally because it has a number of by-products suitable for creating added value [8]. Moreover, it is stated that it can be an alternative to petroleum in the long run due to increases in its use in biodiesel production in recent years. Despite its various uses cotton remains a fiber crop in general. According to 2018 data, cotton constitutes 81% of world fiber production [9].

2. World cotton production

As earlier mentioned, cotton is one of the most important agricultural products widely grown and traded in the world. Approximately 3% of the total cultivation areas of developing countries are devoted to cotton farming.

It is a cash crop with which millions of farmers worldwide can meet the cash needs to satisfy their basic necessities. The income generated from this crop makes great contributions to the welfare level of rural households, especially in developing countries. It is also important for the countries to be an export crop to earn foreign currencies and a raw material for the textile industry [10].

Cotton also provides additional employment opportunities in direct or indirect ways such as textile production, feed, oil, paper, chipboard, and medium density fibreboards (MDF) productions, demand-creating for industrial products such as agricultural machinery-equipment and fertilizers.

Cotton has played an active role in industrialization process in the world since the eighteenth century and continues to play its central role as an important source of income for developing countries today [11].

When having a look at the world cotton production statistics from 1950 to the present; one can observe that production has increased approximately four times despite the limited upward shift in acreage (**Tables 1** and **2**).

The world average seed cotton yield was realized to be 2610 kg per hectare in 2020. China ranked first with an average yield of 9080 kg per hectare among the countries achieving high seed cotton yields in 2020. Although ranking first for many years regarding the seed cotton yield, Australia managed to find second place in the list with an average yield of 5340 kg/ha. Türkiye ranked third on the list with a yield of about 5 tonnes per hectare well above the world average. India, one of the world's largest cotton producers, and the African countries, which have expanded their cotton acreages, grow cotton with yields below the world average [12].

When having a look at the average cotton yields it is noteworthy that there is a significant yield difference among the countries. India, the second largest cotton producer in the world, ranks 37th regarding seed cotton yields with an average of 1380 kg per hectare. It is thought that the basis of these differences is the different ecological conditions and the variations in production techniques of the countries (**Table 1**). In the last 70 years, in parallel with the increases in cotton production in the world, its consumption has increased approximately three times. It is thought that the main factors affecting this increase are the development of the textile and ready-made clothing industry, the increases in the world population and the gross national product per capita, the synthetic fiber consumption trend and prices.

Cotton is grown mostly in Asia. Approximately 69.87% of the world's cotton acreage is in Asia. Amerika, Africa, Europe, and Australia rank second, third, fourth and fifth with the shares of 22.53, 5.81, 1.34 and 0.45% in respective order. In the 20/20 season, approximately 29 million farmers grow cotton in 83 countries worldwide.

No	Country	2016	2017	2018	2019	2020
1	China	475	354	551	681	908
2	Australia	542	415	505	531	534
3	Türkiye	505	489	496	460	494
4	Mexico	467	476	483	442	469
5	Brazil	348	414	431	424	433
6	Bangladesh	346	347	348	338	340
7	Azerbaijan	176	153	176	295	336
8	Kyrgyzstan	314	318	324	328	334
	World	224	213	220	228	261

Table 1.
Seed cotton yields by country between 2016 and 2020 (kg/da).

Rank	Country		2016	2017	2018	2019	2020
1	India	Harvested area	10,830	12,430	12,350	15,029	12,865
		Share (%)	35.91	35.86	37.72	40.93	40.40
2	USA	Harvested area	3848	4492	4130	4699	3521
		Share (%)	12.76	12.96	12.61	12.80	11.06
3	China	Harvested area	2489	2700	2373	2517	2079
		Share (%)	8.25	7.79	7.25	6.86	6.53
4	Pakistan	Harvested area	2489	2700	2373	2517	2079
		Share (%)	8.25	7.79	7.25	6.86	6.
5	Brazil	Harvested area	996	928	1150	1627	1633
		Share (%)	3.30	2.68	3.51	4.43	5.13
6	Uzbekistan	Harvested area	1265	1201	1108	1051	1058
		Share (%)	4.20	3.47	3.38	2.86	3.32
7	Burkina Faso	Harvested area	655	845	473	591	647
		Share (%)	2.17	2.44	1.45	1.61	2.03
8	Benin	Harvested area	419	530	640	670	620
		Share (%)	1.39	1.53	1.95	1.82	1.95
9	Turkmenistan	Harvested area	550	540	535	535	535
		Share (%)	1.82	1.56	1.63	1.46	1.68
10	Türkiye	Harvested area	416	501	519	478	359
		Share (%)	1.38	1.45	1.58	1.30	1.13
	World	Harvested area	30,156	34,666	32,742	36,717	31,840
		Share (%)	100.0	100.0	100.0	100.0	100.0

Table 2.
Cotton harvest area of top 10 producer countries between 2016 and 2020.

About 84% of global cotton production was achieved by 7 countries including India, China, America, Brazil, Pakistan, Türkiye and Uzbekistan [13].

Regarding seed cotton production, the top 10 countries, and their production statistics between 2016 and 2020 were presented in **Table 3**. China alone produced the one-third of the global production of about 83 million tonnes in 2020. Türkiye came 7th with about 2% of global share after Uzbekistan. It is clear from **Table 3** that China, Brazil, Argentina, and Benin increased their seed cotton production successively. World total production followed the same trend except for a slight decrease in 2020. However, seed cotton production of other countries varied across the 5 years and decreased especially in Pakistan, Uzbekistan, and Türkiye in recent years (**Table 3**).

The most important reason for these fluctuations is obviously related to the changes in harvested area (**Table 2**). Yet it might also be arisen due to the variations in fiber length and quality because of the changes in ecological conditions which also alter from place to place by years.

Regarding seed cotton consumption, top 10 countries, and their production statistics available for the latest 5 years between 2015 and 2019 were presented in

No	Countries		2016	2017	2018	2019	2020
1	China	Production	16,029	17,130	18,493	23,505	29,500
		Share (%)	23.75	23.20	25.67	28.03	35.49
2	India	Production	17,308	17,425	14,657	18,558	17,731
		Share (%)	25.65	23.60	20.34	22.13	21.33
3	USA	Production	10,083	12,000	11,133	12,819	9737
		Share (%)	14.94	16.25	15.45	15.29	11.72
4	Brazil	Production	3464	3843	4956	6893	7070
		Share (%)	5.13	5.20	6.88	8.22	8.51
5	Pakistan	Production	5237	5855	4828	4480	3454
		Share (%)	7.76	7.93	6.70	5.34	4.16
6	Uzbekistan	Production	2959	2854	2286	2692	3064
		Share (%)	4.39	3.86	3.17	3.21	3.69
7	Türkiye	Production	2100	2450	2570	2200	1774
		Share (%)	3.11	3.32	3.57	2.62	2.13
8	Argentina	Production	673	616	814	873	1046
		Share (%)	1.00	0.83	1.13	1.04	1.26
9	Burkina Faso	Production	785	844	482	724	783
		Share (%)	1.16	1.14	0.67	0.86	0.94
10	Benin	Production	451	598	678	715	728
		Share (%)	0.67	0.81	0.94	0.85	0.88
	World	Production	67,478	73,842	72,049	83,859	83,113
		Share (%)	100.00	100.00	100.00	100.00	100.00

Table 3
Seed cotton production of top 10 countries between 2016 and 2020 (thousand tonnes).

Table 4. As in the production statistics, the first five countries with the highest seed cotton consumption are the same except for the change in ranking between China and India. Yet, Türkiye comes fifth after Pakistan, moving up two ranks compared to the ranking list for the top 10 largest producer countries.

From the production and consumption statistics presented in **Tables 3** and **4**, it is obvious that the largest cotton-producer countries have also the largest demand for cotton at the same time [14].

World cotton imports show an increasing trend continuously. For example, cotton imports have increased by 66.16% (34,128,689 thousand USD) for the years between 1990 and 2020 [15]. It is obvious from **Table 5** that the commercial value of cotton maintains and increases its importance.

Cotton imports and exports were examined in seed cotton, fiber cotton and cottonseed items separately and the trade values were presented in **Table 5**. World cotton export values show an increasing trend as is the case with import values. According to the export values for the years 1990–2020 cotton exports increased by about 80% [16].

Rank	Country		2015	2016	2017	2018	2019
1	India	Production	5101	5542	5577	4767	6033
		Share (%)	22.77	23.69	21.90	19.17	23.65
2	China	Production	5610	5343	5653	6103	4892
		Share (%)	25.04	22.84	22.20	24.55	19.17
3	USA	Production	2806	3738	4555	4004	4335
		Share (%)	12.53	15.98	17.89	16.10	16.99
4	Brazil	Production	1627	1420	1499	1933	2688
		Share (%)	7.26	6.07	5.89	7.77	10.54
5	Pakistan	Production	1687	1815	2032	1677	1556
		Share (%)	7.53	7.76	7.98	6.75	6.10
6	Türkiye	Production	738	756	882	977	814
		Share (%)	3.29	3.23	3.46	3.93	3.19
7	Uzbekistan	Production	1095	1065	944	769	769
		Share (%)	4.89	4.55	3.71	3.09	3.01
8	Australia	Production	422	551	774	950	414
		Share (%)	1.88	2.35	3.04	3.82	1.62
9	Mexico	Production	205	164	335	414	368
		Share (%)	0.92	0.70	1.32	1.67	1.44
10.	Argentina	Production	303	290	260	266	308
		Share (%)	1.35	1.24	1.02	1.07	1.21
	World	Production	22,402	23,395	25,463	24,864	25,515
		Share (%)	100.00	100.00	100.00	100.00	100.00

Table 4.
Fiber cotton production by country between 2015 and 2019 (in thousand tonnes).

No	Country	Seed cotton import value (thousand $)	Country	Fiber cotton import value (thousand $)	Country	Cotton seed import value (thousand $)
1	China	8,278,844	China	3,562,312	Italia	41,750
2	Vietnam	3,692,816	Vietnam	2,176,156	South Korea	40,876
3	Türkiye	2,543,331	Bangladesh	2,014,804	Saudi Arabia	31,472
4	Pakistan	1,392,682	Türkiye	1,652,640	Japan	27,854
5	Indonesia	1,327,738	Pakistan	1,315,549	Mexico	25,143
6	USA	906,751	Indonesia	774,649	Azerbaijan	7068
7	South Korea	880,825	India	344,649	USA	6507

No	Country	Seed cotton import value (thousand $)	Country	Fiber cotton import value (thousand $)	Country	Cotton seed import value (thousand $)
8	Italia	846,240	Thailand	218,573	Türkiye	3726
9	Germany	755,493	Malaysia	209,468	Spain	3321
10	Egypt	542,087	South Korea	187,109	Zimbabwe	3 02

Table 5.
Cotton imports by country in 2020 (in thousand USD) [15].

The data presented in **Tables 3–5** show that the leading countries in seed cotton exports are more or less the same countries of production and import with slight differences in ranking. However, the ranking differs regarding fiber cotton and cottonseed exports. Yet, USA, India, Türkiye and Brazil can find places among the top 10 countries of seed cotton, fiber cotton and cotton seed exports.

It can be seen from **Table 6** that developing countries must compete with developed countries, such as the USA and China, in order to have a share of the global export market. Countries such as China, India, Brazil and Türkiye play a more active role in the global cotton market as they are more economically developed countries than African and some Asian cotton producer countries.

Cotton, one of the most important industrial crops for China and India, which are the world's largest fiber cotton producers, has 1% share of the GDP in these two countries. It makes up some 0.2% of GDP in Brazil as its share is only 0.03% in Argentina where a shrinkage is experienced in the sector [17]. On the other hand, while cotton

No	Country	Seed cotton export value (thousand $)	Country	Fiber cotton export value (thousand $)	Country	Cotton seed export value (thousand $)
1	China	10,998,837	USA	5,969,433	USA	94,576
2	USA	7,013,425	Brazil	3,226,916	Greece	69,875
3	India	5,809,190	India	1,448,516	Ivory Coast	15,800
4	Brazil	3,319,722	Greece	449,992	Benin	12,491
5	Vietnam	2,707,568	Benin	448,829	Ethiopia	9241
6	Pakistan	2,642,547	Australia	307,109	Türkiye	8320
7	Türkiye	1,463,452	Burkina Faso	263,296	Brazil	6526
8	Uzbekistan	1,200,763	Egypt	161,604	Israel	4592
9	Italia	961,167	Türkiye	159,811	India	3338
10	Germany	694,732	Uzbekistan	147,318	Zambia	3209

Table 6.
Cotton export values by country as of 2020 (thousand USD) [15].

production in Brazil constitutes 0.2% of GDP, it accounts for only 0.03% of GDP in Argentina, where the sector is shrinking.

The countries having a large share of global production such as China and USA and their production and stock sizes shape international market prices for cotton [18]. The production volumes and prices of other substitute fiber products, that are important for the textile industry such as wool, mohair, synthetic yarn, and the value of the USD against other currencies also play an important role in formation of cotton prices at international market [19]. In addition, the subsidies given by developed countries to their cotton producers negatively affect the competitiveness of other countries [20].

3. Cotton in Türkiye

Türkiye has the advantages of great agricultural potential and diversity. For that reason, the export-creating effect of production increase in agriculture is greater than that in industry [21]. Again, the driving force in cotton exports of Türkiye is the agriculture-based industries. It is a strategic crop for Türkiye and so it is called 'white gold' because of its value-added capacity, rather than its color.

Extension of cotton production in Türkiye was initiated by the events in remote countries. For example, cotton production in the USA, ranking first in global cotton production, decreased and prices increased due to the civil war started in 1861, This event caused a bottleneck for the UK, the most important cotton importer of that time due to the great Industrial Revolution, This bottleneck, of course, caused the British textile industrialists fall into a difficult position, Upon this, the organization of these industrialists, Manchester Cotton Supply Association, MCSA, started to search and develop new cotton production areas.

Adana[1] was also one of the areas found to meet the cotton demand. Upon with the cotton demand of UK, the Turkish government of that time in İstanbul issued various incentives to increase cotton production in Adana in 1862 [22]. In this context, it was decided to provide tax relief to farmers for 5 years [23]. Moreover, Sultan issued an edict to provide cotton seeds for the farmers and 165 thousand okkas[2] of seeds were brought from Egypt and distributed to the farmers. Again, in addition to all these incentives, UK sold 6000 tons of cotton seeds to the landowners in 1864 [24, 25].

The widespread cotton production in Adana inevitably led to the establishment of the textile industry. Until the 1860s, cotton cleaning and pressing processes were carried out with extremely primitive methods. However, upon the Ottoman Empire made cotton production an official policy, there had been a great increase in the investments of foreign states and important factories were established [26]. As a result, an increase in cotton production from 10 million kilograms to 50 million kilograms was achieved, which caused Adana to develop in agriculture, industry and trade.

Accordingly, it was necessary to take steps to make urban life more modern. However, by the end of the American Civil War, the European states turned to cheaper American cotton, which interrupted the development of Turkish cotton. Nevertheless, this experience formed an important basis for the Republican period. During the Republican period, in Türkiye, the main cotton production areas were

[1] Adana is a province and the sixth largest metropolitan area in Türkiye, located on the Mediterranean coast in the south of the Anatolian peninsula. It is one of the country's leading agricultural, commercial and cultural centres.

[2] Okka was a unit of weight used in Türkiye in the past, equal to 1283 kg in metric system.

considered. The construction of irrigation facilities in the Sakarya Valley to make use of the Sakarya river and Porsuk stream, enabled the development of cotton farming in this region [27]. In this period, the experience of foreign experts was also used in the development of cotton farming. While the production increase attempts continued on the one hand, the establishment of cotton plants formed the basis of Turkish Industry sector [28]. The incentives provided for the expansion of cotton production, the provision of infrastructure services, and product purchase guarantees, along with the establishment of state factories, Türkiye has reached its current state in cotton production and processing.

Turkish cotton acreage for the 20-year period between 2001 and 2020 is given in **Table 7**. A decrease of about 47% was recorded during this period, more than the decrease in world cotton acreage in percentage. As a result, although fluctuated throughout this period, the share of Türkiye in world cotton acreage decreased to about 1% from about 2% and cotton cultivation area is 1.13%. It is seen that cotton cultivation areas in Türkiye are fluctuating and decreasing.

Years	Türkiye		World		Share of Türkiye (%)
	Acreage (Ha)	Index (2000 = 100)	Acreage (Ha)	Index (2000 = 100)	
2001	684,665	100.00	34,624 130	100.00	1.98
2002	721,077	105.32	30,751 267	88.81	2.34
2003	637,329	93.09	30,996 622	89.52	2.06
2004	640,045	93.48	35,018 796	101.14	1.83
2005	546,040	79.75	34,749 423	100.36	1.57
2006	589,700	86.13	34,437 460	99.46	1.71
2007	529,853	77.39	33,500 005	96.75	1.58
2008	494,891	72.28	31,035 321	89.63	1.59
2009	419,873	61.33	30,221 985	87.29	1.39
2010	480,439	70.17	31,795 011	91.82	1.51
2011	541,952	79.16	34,490 945	99.61	1.57
2012	488,496	71.35	34,691 598	100.19	1.41
2013	450,890	65.86	32,219 726	93.06	1.40
2014	466,839	68.19	34,613 269	99.97	1.35
2015	434,000	63.39	31,623 309	91.33	1.37
2016	416,002	60.76	30,155 719	87.09	1.38
2017	501,478	73.24	34,666 094	100.12	1.45
2018	518,634	75.75	32,741 708	94.56	1.58
2019	477,807	69.79	36,716 883	106.04	1.30
2020	359,220	52.47	31,840 226	91.96	1.13

Source: www.fao.org

Table 7.
Cotton acreage of Türkiye and its share in total world cotton acreage.

Having opened to irrigation by the implementation of the Southeastern Anatolia Project (GAP), the Southeastern region has turned into new cotton production areas. The provinces with the highest cotton acreages are presented in **Table 8**. According to the 2021 statistics, Şanlıurfa, Diyarbakır, Aydın, Hatay, Adana and İzmir provinces have 86% of Turkish cotton acreage [29]. Even the southeastern provinces of Türkiye, Şanlıurfa and Diyarbakır, rank first and second in cotton acreage respectively. Adana province where cotton production first started in Türkiye, only managed to find 6th place in the list.

According to 2021 statistics seed cotton production in Türkiye was 2250 thousand tonnes. The shares of the leading provinces in cotton production are given in **Table 9**. Nearly 60% of seed cotton production is achieved in the Southeastern Anatolia region (**Table 9**). Adana province, which once produced more than half of Türkiye's cotton production, ranks sixth in production as in acreage due to that it has now turned to different agricultural products [16].

While having a share of about 1% in world cotton exports Türkiye's share in world cotton imports is quite high with more than 12%. Whereas Türkiye was an important exporter in meeting the world's cotton demand. However, the rapid development of the Turkish textile industry has led to an increase in the demand for cotton.

Rank	Province		2017	2018	2019	2020	2021
1	Şanlıurfa	Acreage	2237	2314	2088	1287	1835
		Share (%)	44.60	44.62	43.70	35.84	42.44
2	Diyarbakır	Acreage	428	480	477	404	552
		Share (%)	8.53	9.26	9.98	11.24	12.78
3	Aydın	Acreage	646	537	465	545	493
		Share (%)	12.88	10.35	9.72	15.16	11.41
4	Hatay	Acreage	518	485	456	332	390
		Share (%)	10.33	9.36	9.55	9.23	9.01
5	İzmir	Acreage	274	277	253	277	262
		Share (%)	5.47	5.35	5.29	7.72	6.07
6	Adana	Acreage	318	363	375	230	219
		Share (%)	6.34	6.99	7.84	6.40	5.06
7	Manisa	Acreage	59	103	90	92	124
		Share (%)	1.17	1.98	1.89	2.55	2.87
8	Denizli	Acreage	91	84	72	95	97
		Share (%)	1.82	1.63	1.51	2.66	2.24
9	Adıyaman	Acreage	66	80	72	55	61
		Share (%)	1.32	1.54	1.51	1.54	1.42
10	Mardin	Acreage	84	108	130	74	60
		Share (%)	1.67	2.08	2.72	2.07	1.38
	Türkiye	Acreage	5015	5186	4778	3592	4323
		Share (%)	100.00	100.00	100.00	100.00	100.00

Table 8.
Cotton acreage by province between 2017 and 2021 (in thousand decares).

Rank	Province		2017	2018	2019	2020	2021
1	Şanlıurfa	Production	1028	1028	813	567	893
		Share (%)	41.97	39.99	36.97	31.98	39.68
2	Diyarbakır	Production	217	244	234	218	309
		Share (%)	8.87	9.51	10.62	12.27	13.74
3	Aydın	Production	331	279	246	272	265
		Share (%)	13.52	10.87	11.20	15.31	11.76
4	Hatay	Production	266	264	220	175	210
		Share (%)	10.84	10.27	9.98	9.86	9.34
5	İzmir	Production	144	156	142	149	151
		Share (%)	5.86	6.07	6.45	8.39	6.69
6	Adana	Production	168	206	206	123	114
		Share (%)	6.87	8.02	9.35	6.93	5.05
7	Manisa	Production	32	61	55	56	73
		Share (%)	1.30	2.38	2.50	3.13	3.25
8	Denizli	Production	47	43	38	50	53
		Share (%)	1.94	1.65	1.74	2.82	2.34
9	Mardin	Production	42	57	65	40	33
		Share (%)	1.71	2.21	2.97	2.24	1.45
10	Adıyaman	Production	32	41	35	27	31
		Share (%)	1.31	1.58	1.59	1.52	1.37
	Türkiye	Production	2450	2570	2200	1774	2250
		Share (%)	100.00	100.00	100.00	100.00	100.00

Table 9.
Unseeded cotton production amount by province between 2017 and 2021 (thousand tonnes).

Consequently, Türkiye has become one of the important cotton importers due to insufficient production to meet the domestic demand [29].

Seed cotton export and import quantities of Türkiye were about 136 thousand tonnes and 1200 thousand tonnes respectively in the year 2021. Turkish seed cotton exports and imports are presented by countries in **Figure 1** [30].

With the increased awareness of the non-biodegradability of synthetic fibers, their damage to the environment, and the augmented importance of human and environmental health especially with the effect of COVID-19 pandemic the share of cotton in the world market has been increasing day by day [29].

Although having been a net seed cotton importer in recent years, Türkiye gained multi-benefits from processing the imported seed cotton into yarn for textile industry, oil for human nutrition and pulp for animal nutirition. Thus, cotton import meets the demand of the textile and ready-made clothing manufacturing industry, which brings about growth of the sector, resulting in an increase in employment and export revenues [31].

Both domestic seed cotton production and imported seed cotton are utilized to create added value in the textile and ready-made clothing industry.

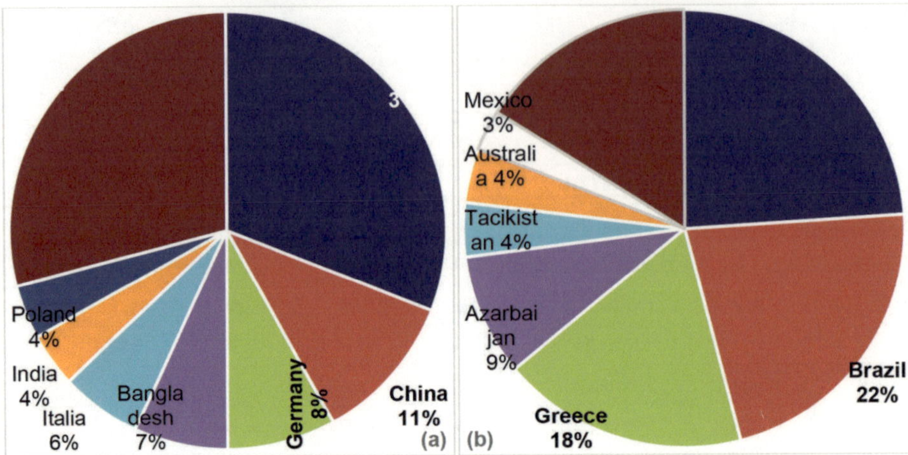

Figure 1.
Cotton export (a) and import (b) of Türkiye by country in 2021.

Türkiye's textiles imports and exports were about 7.8 billion dollars and 12.3 billion dollars respectively in 2020. Ready-to-wear exports were about 15 billion dollars, as the imports were around 1.3 billion dollars. When we consider the total export and import values including textile, ready-made clothing and leather products Türkiye had a net foreign trade surplus of about 18 billion dollars in this sector (**Table 10**).

The development course of the textile, ready-made clothing industry had an increasing trend in general, except the years between 2008 and 2010. The global economic crisis that started in the USA affected Turkish Agriculture parallel to its worldwide harmful effects.

Since agricultural products are the basic consumables, they are compulsorily demanded by consumers. In addition, the effects of the crisis were felt less in this sector since they are less sensitive to changes in supply and demand. The shrinkage in demand for agricultural products during crisis is less than that for industrial goods [31].

Therefore, the effects of the crisis were felt less in agriculture-based industries, compared to other industries.

It is thought that cotton will continue to be one of Türkiye's most strategic products as long as we cannot steer the employment created by the textile and ready-to-wear industry to more profitable areas and produce a more value-added agricultural product than cotton. Cotton is an agricultural crop with no alternative for Türkiye and accordingly, cotton-related policies should be in line with this strategic

Sectors	Exports	Imports	Foreign Trade Balance
Textile	12,342	7772	4570
Ready-made clothing	14,988	1293	13,695
Total	27,330	9065	18,265

Source: TRADEMAP [30].

Table 10.
Turkish textile and ready-made clothing foreign trade values for the year 2020 (in million dollars).

importance in Türkiye. The economic cost of the abandonment of cotton farming will definitely be much higher than the burden on the budget due to the support payments and the subsidies for sustainable cotton production today. The employment created in agriculture and industry, the main and by-products, and the positive effect on the balance of payments make cotton indispensable for Türkiye.

In the Turkish National Cotton Industry Report it was declared that "627 thousand tonnes of fiber cotton and 883 thousand tonnes of cottonseed were obtained from 2.2 million tonnes of cotton seed production in the 2019/20 season. In addition, 133 thousand tonnes of edible oil and 618 thousand tonnes of fodder pulp were obtained from the processing of cottonseed as well as 44 thousand tonnes of linter and 96 thousand tonnes of cotton waste were made available to many industries, especially medical supplies and cosmetics. In this sense, when only the main product, 'fiber cotton', is considered, it will be obvious that through processing in textile and ready-made clothing industries domestic cotton production, which is worth 1 billion 204 million dollars, brought to national economy with an increase 11 times in gross and 2.5 times in net value. That almost half (45.17%) was obtained in foreign currency increases the importance of this earning even more.

With the inclusion of sustainability, environmental, social and economic dimensions among the priorities of the cotton industry, cotton, which is a basic and indispensable raw material and natural fiber, has become more important for the textile and ready-made clothing industries. With its easy biodegradability property due to being a cellulose-based natural fiber, cotton has an important advantage especially when compared to polyester and other synthetic fibers. In addition to that Türkiye has also an important place in the use of sustainable cotton production methods, such as organic and 'Better' cotton production, even conventionally grown cotton has also sustainability since Turkish cotton is non-GMO.

4. Conclusion

The fluctuations in the world cotton prices and the easy import of fiber cotton affect the cotton farmers in Türkiye negatively. This causes farmers to seek alternative crops and shrink their cotton-growing areas. Due to its importance, cotton is a product especially supported by producer countries in the world. Thus, according to International Cotton Advisory Committee (ICAC) data, government support have reached high levels in recent years. In order to increase the production of cotton, also called "white gold", alternative supports, such as price support, customs measures and crop insurance support should be considered. In general, it would be beneficial to reorganize cotton policies in order to solve high-cost problems, increase production and create a competitive environment. It would be fairer to determine cotton prices not by a few countries such as the USA and China but by the contributions of other producer countries on a global scale. In addition, along with sharing the knowledge markets should be freed from the monopoly of GMO cotton and scientific studies should be supported more for the development and extension of high-yield cotton cultivars to increase the quality of cotton worldwide.

Conflict of interest

The authors declare no conflict of interest.

Author details

Songül Akın
Department of Agricultural Economy, Dicle University, Diyarbakır, Türkiye

*Address all correspondence to: sakin@dicle.edu.tr

References

[1] Özertan G. Yeni Tarım Düzeni ve Türkiye Tarım Sektöründe Kalkınma İçin Teknoloji Kullanımının Rolü. Kalkınmada Yeni Yaklaşımlar, Yayına Hazırlayanlar: Ahmet Faruk Aysan, Devrim Dumludağ, İmge Kitabevi Yayınları, Ankara içinde. 2014. pp. 209-242

[2] Tufan A. Türkiye Ekonomisi. Ankara Üniversitesi Ziraat Fakültesi Yayınları. 1997. p. 1487. Available from: http://hdl.handle.net/20.500.12575/17469

[3] Küçük N, Veissi S. Pamuk Üretiminin Stratejik Önemi Üzerine Genel Bir Değerlendirme. TURAN Stratejik Araştırmalar Merkezi. 2019;**11**(44):391-398

[4] Aydogdu MH, Karli B, Parlakci Dogan H, Sevinc G, Eren ME, Kucuk N. Economic analysis of agricultural water usage efficiency in the GAP-Harran plain: Cotton production sampling, Sanliurfa-Turkey. International Journal of Advances in Agriculture Sciences. 2018;**3**(12):12-19. ISSN: 2456-7515

[5] Gürel A, Akdemir H, Emiroğlu ŞH, Kadioğlu H, Karadayi HB. Türkiye Lif Bitkileri (Pamuk Tarımı ve Teknolojisine Genel Bakış ve Diğer Lif Bitkileri). In: Türkiye Ziraat Mühendisliği V. Ankara, Türkiye: Teknik Kongresi; 2000. pp. 525-566

[6] Doğan S. Farklı Pamuk (Gossypium Spp.) Genotiplerinin Tohum Özellikleri, Yağ Oranı ve Gossypol İçerikleri Yönünden Değerlendirilmesi. Kahramanmaraş: Kahramanmaraş Sütçü İmam Üniversitesi, Fen Bilimleri Enstitüsü, Yüksek Lisans Tezi; 2021. pp. 1-62

[7] Oğuz FK. Değerini Bilmediğimiz Bir Ürün: Pamuk Tohumu. Türkiye Yem Sanayiciler Birliği, Yem Mağazin Dergisi. 2006;**43**:47-52

[8] Zahedi M, Eshghizade HR, Mondani F. Energi use efficiency and economical analysis in cotton production system in an arid region: A case study for Isfahan Province, Iran. International Journal of Energi Economics and Policy. 2014;**4**(1):43-52

[9] Yakti Ö. Amerikan İç Savaşi Ve Adana: Pamuk Tariminin Adana'nin Modernleşme Sürecine Etkisi Ankara Üniversitesi Türk İnkılâp Tarihi Enstitüsü Atatürk Yolu Dergisi S 54. Bahar. 2013;**2013**:273-296

[10] Fortucci P. The Contributions of Cotton to Economy and Food Security in Developing Countries. FAO; Jul 2002. pp. 1-10. Available from: https://staging.icac.org/Meetings/cgtn_conf/documents/11_fortucci.pdf

[11] Anonymous. FAO & ICAC Measuring Sustainability in Cotton Farming Systems Towards a Guidance Framework, ICAC Report: III-V. Expert Panel on Social, Environmental and Economic Performance of Cotton Production with the FAO Plant Production and Protection Division. Rome, Italy; 2015. pp. 1-107. Available from: https://www.fao.org/3/i4170f/i4170f.pdf

[12] Raporu P. T.C. Ticaret Bakanlığı Esnaf, Sanatkârlar ve Kooperatifçilik Genel Müdürlüğü, 2019 Yılı Pamuk Raporu, Nisan 2020. 2020. pp. 1-40. Available from: http://www.upk.org.tr/User_Files/editor/file/2019%20Pamuk%20Raporu.pdf

[13] ICAC. International Cotton Advisory Committee. Washington DC, USA: Cotton Data Book 2020; 2020

[14] ICAC. International Cotton Advisory Committee. Washington DC, USA: Cotton This Month; 2022

[15] UN Comtrade. Database. 2022. Available from: https://comtrade.un.org/data [Accessed: March 23, 2023]

[16] Khalili MY. The impact of support policies on cotton production in turkey and future forecast [MSc Thesis]. Harran University Graduate School of Natural and Applied Sciences Department of Agricultural Economics. Şanlıurfa, Türkiye; 2023

[17] Gillson I, Poulton C, Balcombe K, Page S. Understanding the impact of cotton subsidies on developing countries. Munich Personal RePEc Archive. 2004:1-144. Available from: https://mpra.ub.uni-muenchen.de/15373/1/MPRA_paper_15373.pdf

[18] Koondhar MA, Qiu L, Liu X, Abbassi F, Koondhar MA, Ge H. Effects of market price, cultivating area and price regulation on cotton production in China. African Journal of Agricultural Research. 2018;**13**(16):858-865

[19] Liu B, Hudson D. A preliminary analysis of the effects of China's cotton tariff on the Chinese and US cotton markets. Estey Journal of International Law and Trade Policy. 2019;**20**(1753-2019-2986):15-27

[20] Tokel D. Dünya Pamuk Tarımı ve Ekonomiye Katkısı. MANAS Journal of Social Studies. 2021;**10**(2):1022-1037

[21] Tüsiad. Tarıma Dayalı Sanayide Durum. İstanbul: Türk Sanayicileri ve İş Adamları Derneği, Yayın No: TÜSIAD-T/87.1.96; 1987

[22] Deane P. İlk Sanayi İnkılâbı (Çev. Tevfik Güran). Ankara: TTK Basımevi; 2000

[23] Aricanli T. 19. Yüzyılda Anadolu'da Mülkiyet, Toprak ve Emek. In: Keyder Ç, Tabak F, editors, (Çev. Zeynep Altok) Osmanlı'da Toprak Mülkiyeti ve Ticari Tarım. İstanbul: Tarih Vakfı Yurt Yayınları; 2010

[24] Toksöz M. Bölgesel Ekonominin Oluşumu Göçebe Hayattan Ticari Tarıma Çukurova. Toplumsal Tarih Dergisi. 2009:191

[25] Townsend T. 1B - World natural fibre production and employment. In: Handbook of Natural Fibres. 2nd ed. Cambridge, UK: Woodhead Publishing; 2020. pp. 15-36

[26] Doğan O-EFÖ. Çukurova'ya Bereket Getiren Projeler. Çamlıca Yayınları: İstanbul; 2011

[27] Doğukan SR. Türkiye'de Pamuk. İktisadi Yürüyüş Dergisi. 1941;**3**(29):12-13

[28] Yurtoğlu N. Cotton Productıon And Cotton Polıcıes Durıng Ataturk Era in Turkey (1923-1938). Tarih Araştırmaları Dergisi. 2019;**39**(68):504-554

[29] Aydin Can B. Dünyada ve Türkiyede Pamuk Üretimi ve Dış Ticaretindeki Gelişmeler. İksat Yayınevi, Ankara: Tarımsal Üretim Ve Pazarlamada Akademik Yaklaşımlar; 2022

[30] TRADEMAP. 2022. Available from: https://www.trademap.org/

[31] Karli B, Kadakoğlu B, Gül M. Dünya ve Türkiye'de Pamuk Üretimi ve Dış Ticaret Yapısı. In: V. International Multidisciplinary Congress of Eurasia, Temmuz 24-26; Barcelona, İspanya. 2018. pp. 129-136